D1388413

Field Guide to
Mushrooms
and other Fungi
of Britain and Europe

How to identify fungi

> ➤ **Pore fungi**
> **Pages 26 – 45**
>
> These have a stem and tubes ending in pores. They include the Cep, Bay Bolete, Orange Birch Bolete and the Yellow Cracked Bolete.
> (Left: Larch Bolete)

> ➤ **Gill fungi,**
> **stem with ring**
> **Pages 46 – 67**
>
> These fungi have a ringed stem and gills, for example the Field Mushroom, Fly Agaric and the Death Cap.
> (Left: Golden Bootleg)

➤ Warning

Even the best nature guides can only help you identify fungi with complete certainty if you follow and identify each and every one of the distinguishing features and identification guidelines for the specific fungus. If you are slightly uncertain about just one of the distinguishing features, then you cannot safely say that the mushroom in front of you is the same mushroom as described in the book. If you have the slightest doubt about which species of mushroom you have collected, you should take your find to a mushroom expert.

For ease of identification, fungi have been divided into four colour groups

➤ **Gill fungi,
stem without ring
Pages 68 – 137**

These fungi have a ringless
stem. Examples include the
Milk Caps, Russulas
and Inocybes.
(Right: Saffron
Milk Cap)

➤ **Other fungi
Pages 138 – 185**

These fungi take a variety of
shapes, such as the Tinder
Fungus, Common Earthball,
Jelly Ear, and
Devil's Fingers.
(Right: Devil's
Fingers)

SYMBOLS:

Each description of a fungus
also contains a symbol to
indicate whether it is edible,
inedible or poisonous. Unless
stated otherwise, the same is
true for similar species.

deadly poisonous	
poisonous	
inedible	
edible with caution	
edible	

Step-by-step identification

Slippery Jack I
Suillus luteus

APPEARANCE: Cap is hemispherical, chocolate or orange-brown ①, and up to 10cm wide; very greasy in damp weather. In young specimens, remnants of the veil often cling to the cap ③; Tubes are yellow ④, olive-yellow on older mushrooms, with very narrow pores. Stem is normally cylindrical, yellow near the top, with dark spotting; below the prominent ring, the stem has ? coating.

FLESH: The whit? flesh ② does not discolour when c? ?mell and taste are pleasant and mild.

DISTRIBUTION: Exclusively under pine trees.

EDIBILITY: Young specimens are excellent for cooking, but older mushrooms are very soft. The skin of the cap should be removed before cooking. In recent years there have been reports that repeated consumption of large quantities of Slippery Jack can cause allergic reactions.

TYPICAL FEATURES
Slippery Jack is only found under pines and can be recognised by its dark brown, greasy cap, a ? ?nent ring o? ?m.

40   41

Step 1:
COMPARE WITH THE MAIN PICTURE

Each main picture shows the characteristic features of the fungus in its natural surroundings. Easily confused species are often shown on the same page.

Step 2:
IDENTIFICATION OF DISTINGUISHING FEATURES

The two illustrations and additional photograph clearly show the distinguishing features and provide useful additional information for identification purposes. Using these pictures will enable you to identify a species with certainty. In many cases, similar species are also pictured, to help avoid mistaken identification.

Step 3: Identification text

The identification text clearly describes the most important classification features of the fungi. Numbered photographs and graphics illustrate the most important features. There is also information about the distribution and habitat of the fungus. This can often be a great help in identifying individual species. Finally, the identification text also offers information on whether a specimen is edible or poisonous. The text often finishes with information and warnings about other species of fungus with a similar appearance.

Step 4: Calendar clock

The time of year when you spot a mushroom growing is also a great help to identification. The grey segments indicate the main fruiting period when you will most likely find these mushrooms. Of course, variations in climate from area to area and year to year mean that they may be subject to slight alteration.

Time of year when the fruiting bodies appear

Step 5: Infobox

The coloured Infobox provides important additional information about distinguishing or typical features. Together, these steps can help you to identify the desired species with accuracy.

TYPICAL FEATURES
Slippery Jack is only found under pines and can be recognised by its dark brown, greasy cap, and prominent ring on the stem.

Fungi

> Golden Bootleg caps open very wide when ripe

The fascinating world of fungi

The world of fungi is astonishing in its variety. In addition to the well-known species of fungi, there are a number of very interesting, lesser known species. Most people are mainly interested in the wild mushrooms found from spring to autumn in woods and meadows, whose tasty caps can be turned into delicious dishes.

What are fungi?

Unlike flowering plants, fungi do not contain chlorophyll and so cannot live by photosynthesis. Instead, they live by breaking down organic matter in the environment; in doing so they introduce life-giving nutrients to the ecosystem. Fungi are alongside plant and animals in a separate phyllum

or order of classification called 'the fungi'.

> Common Puffball

> Orange Peel Fungus

An introduction to mycology

Most fungi grow in the soil or on wood and other plant material, although some are parasites on other fungi. They consist of individual threads (hyphae), which together form the dense, white structure called the 'mycelium'. What we casually refer to as a mushroom is only the 'fruiting body', which grows from the mycelium, under the right climatic conditions. The fruiting body produces vast numbers of spores and this is how the fungus reproduces.

The range of species

There are several thousand species of fungus in Europe. Many of them are so similar that even an expert can only distinguish between them by using a microscope. This book describes 160

> Jelly Antler Fungus

species, and mentions a further 100 similar species and their important characteristics, which are identifiable without the need for a microscope. Some of them do not grow in Great Britain.

> Rosy Earthstar

> Common Morel

SCIENTIFIC NAMES

The scientific name of a fungus is the only sure way to identify it. Until recently, very few fungi had common names in English and those that did may have had several names depending on the area in which they were found. Some fungi have two scientific names, which only adds to the confusion. In some cases we have given the best-known name, rather than the newest name, in order to help the reader.

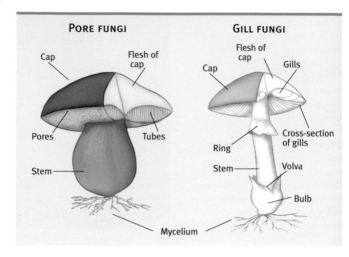

PORE FUNGI

Cap

Flesh of cap

Pores

Tubes

Stem

GILL FUNGI

Flesh of cap

Cap

Gills

Cross-section of gills

Ring

Stem

Volva

Bulb

Mycelium

Shapes of fungi

Fungi can be found in a great number of different shapes and colours. There are pore fungi and gill fungi, tooth fungi and club fungi, funnel caps, bracket fungi, jelly fungi and cup fungi. Most fungi, however, have a simple and similar anatomy: a stem that is anchored in the soil topped by a cap, on the underside of which the characteristic spore-bearing layer can be found.

Boletes

Boletes or pore fungi have a fruiting layer beneath the cap, which consists of numerous tubes that can be easily peeled away from the cap flesh. The openings to these tubes are called pores; in some species they are brightly coloured, and in others they are so tightly

packed that they can hardly be distinguished with the naked eye. In other species the pores are wide and well-spaced.

Gill fungi with a ring

Many of our favourite species of fungi, such as the Field Mushroom and the Parasol Mushroom, are gill fungi with a ringed stem, but some of the most poisonous species, such as the Death Cap, belong to the same category. The ring can sometimes drop off the stem, so you should always look for other distinguishing features.

Gill fungi without a ring

Most gill fungi have no ring and this is the largest group of fungi in the book. Some, such as the Milk Caps, are easy to identify (because of the milk), but others are hard to identify, such as the Brittle Gills and Inocybes.

Other shapes of fungi

Many fungi - from the Morels to Cup Fungi or Tooth Fungi - have fruiting bodies that do not display the characteristics of the classic Mushroom anatomy – a stem and fleshy cap with gills or pores underneath. Other shapes include the folds and ribs of the Chanterelle, or the coral-like shape of the Urchin of the Woods. These fungi are grouped together in the final chapter of the book, along with the bracket fungi that grow on trees or on dead wood.

The volva

An important distinguishing feature of gill fungi is the volva, a membranous cup that covers the base of the stem in several species of the genus Agaricus, including the deadly poisonous Death Cap (see the section on gill fungi, p10). When collecting mushrooms, always be sure to remove the whole stem from the soil, so you can see whether or not they have a volva.

> Bay Bolete is a pore fungus

> Contrary Web Cap has no ring on the stem

> The Field Mushroom has gills and a ringed stem

> Scarlet Elf Cup is unmistakable

> The gills of Red Edge Bonnet are red in cross-section

Caps, pores and gills

The cap is not only the most important edible part of the fruiting body, it also plays a key role in the identification of fungi. The underside of the cap contains the layer that holds the spores. In most fungi, it takes the form of either pores or gills.

Different shapes

The mushroom cap, and the whole of the fruiting body, may take several different forms. The cap often changes shape while growing, from rounded or spherical to umbrella-like or other shapes. Even fully developed mushrooms of the same species can look very different, depending on their habitat and the weather conditions.When identifying fungi, examine as many fruiting bodies as possible.

Tubes and pores

Fungi with a fruiting layer below the cap that consists of tubes and pores are especially good to eat. The Cep and related edible species are all pore fungi. The layer of tubes can be peeled away from the cap. Typical characteristics, for example the colour of the tubes in cross-section, the colour of the pores and the size of the pores, are given in the identification text for each species. The layer of tubes may be slightly concave or it may be closely attached to the point where it joins the stem.

The gills

The gills underneath the cap are thin blades that radiate outwards from the central stem to the edge of the cap, just like the spokes of a bicycle wheel. They may change colour as the mushroom ages. The gills are often flexible but may also be brittle. The gills can be very tightly packed or widely spaced. Gill cross-sections may have different shapes, have different colours, be shiny or fluffy. It is very important when identifying species of gill fungi to note how the gills attach to the stem. In some fungi, the gills do not touch the stem at all, in which case they are said to be 'free'. In others, the gills may lie flat against the stem (adnate), or be partially detached (adnexed).

> Pores on the Cep

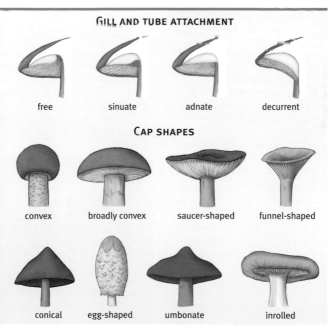

GILL AND TUBE ATTACHMENT

| free | sinuate | adnate | decurrent |

CAP SHAPES

| convex | broadly convex | saucer-shaped | funnel-shaped |

| conical | egg-shaped | umbonate | inrolled |

> The skin on the cap of Porcelain Fungus is slimy

External appearance of fungi

The caps and stems of diverse fungi can look very different. The caps may be scaly or cracked, or have a fluffy or felt-like surface. The caps of some fungi may be smooth, shiny, or even greasy. The texture of the surface of the stem can also be of great help when trying to identify a species of fungus.

Changing colours

The colour of the skin of the cap is not necessarily a definite distinguishing feature. That is because it can change, even quite dramatically, depending on the stage of maturity and the weather conditions. The flesh of some fungi absorb a lot of water (when they are said to be 'hygrophorous'), giving the caps a darker appearance, but they can turn much paler in dry weather. In other species, the cap may appear to be ribbed in wet weather, due to the cap becoming translucent.

Veil

Young fruiting bodies of some varieties of fungus are covered in a

> Split Fibrecap has a fissured cap when mature

veil, which may be tough, stringy or fluffy. As the fungus grows, the veil tears, but remnants may remain on the cap, stem or at the base of the stem. The white spots on the cap of Fly Agaric are an example of this, as is the volva of the Death Cap. Veil remnants may help to identify a species, but they can be absent, for instance if they have been washed away by rain.

The ring

The ring is another remnant of the veil which covers the fruit layer of some young fungi. As the cap opens out, the veil breaks at the edge of the cap, and the ring is left hanging on the stem. In dry weather, the ring may even remain attached to the edge of the cap, and in older mushrooms it may be entirely

> Parasol Mushrooms with scaly caps

absent. To avoid making dangerous errors, examine a number of fruiting bodies to see if any have a veil, volva or ring.

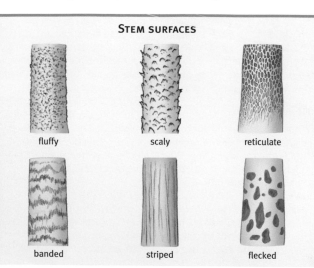

STEM SURFACES

fluffy

scaly

reticulate

banded

striped

flecked

> A group of fungi grow in a 'fairy ring'

Growing behaviour of fungi

Fungi can be divided into three groups, depending on the way they grow. They are 'mycorrhizal fungi', which form a symbiotic relationship with a tree or plant, 'saprophytic fungi' that live on dead plant matter and help to decompose it, and 'parasitic fungi', which can attack plants and kill them.

Just dry theory?

Knowledge of how a fungus lives and grows is not merely academic when it comes to identifying species, because many fungi can only be found under or near a particular species of tree. For instance, the Brown Birch Bolete is only found under birch trees, the Orange Oak Bolete under oaks and Tinder Fungus on dead or dying beeches and birches.

Mycorrhizal fungi

Mycorrhizal fungi form a symbiotic relationship with certain species of tree, enabling the tree roots and the mycelium to exchange organic substances in a way that benefits both organisms. Many popular edible mushrooms, such as almost all the Boletes, Brittle Gills and Milk Caps, belong to this category. Some can only be found under a particular species of tree; others can form this symbiotic partnership with several species of tree.

Saprophytes

Saprophytes live on dead organic material, plant remains and dead wood, or on animal remains. Without these fungi and the decomposing bacteria that replace them when the substrate is sufficiently broken down, nature would drown in its own waste. Examples of this type of fungus are the widespread Mottlegills, such as the Brown Mottlegill, which grow in large numbers in heavily fertilised fields and on rotted cow dung.

Parasites

Parasites take all the nutrients they need for survival from a living host. Of the fungi in this book, most parasitic fungi have a tree as their host, and over time, the tree may be seriously damaged or even die. When identifying parasites, it is important to note whether the fungi is growing on wood that is still living or on dead wood. Many parasitic fungi, such as Root Fomes, are much-feared because of the damage they cause to timber.

Fairy rings

Some species of fungus, including the Wood Blewit, the St George's Mushroom and, of course, the Fairy Ring Mushroom, regularly form what is known as 'fairy rings'. These consist of a number of fruiting bodies growing in a ring to form a circle. Once upon a time, it was believed that witches met inside the rings and fairies danced in them. In fact, it is the result of a mycelium radiating out in all directions from a central point, and forming fruiting bodies at its extremities. Fairy rings may be irregular or broken, or consist of only a few fungi.

> The partnership between Ceps and spruce has advantages for both fungus and tree

> Sulphur Tuft is a saprophyte that lives on dead wood

> This Sulphur Polypore is a parasite on this living tree but it can also live as a saprophyte

Fungi and their environment

If you want to find or collect different species of mushroom, you need to know where to find them, because fungi always live in harmony with their environment. Many popular species of edible mushroom live in partnership with trees. In addition, they are only found beneath particular species of tree, in grassland or where trees grow sparsely. Four different types of woodland are described here, each with a characteristic species of fungus, but the characteristics are representative of many other types of habitat.

Look at the trees

A survey of the surrounding trees is very useful for identifying many species. In some cases, this can help to distinguish between species with a very similar appearance. For example, the Brown Birch Bolete is easily confused with the Grey Bolete (*Leccinum griseum*), but the latter is only found under hornbeams and not under birches. The Larch Bolete grows under larches, but Slippery Jack can only be found under pine trees.

Living on wood

Like mycorrhizal fungi, many saprophytic and parasitic fungi are linked to various habitats or species of tree. The edible Two-toned Pholiota mainly grows on dead deciduous wood, and the

SPRUCE WOODLAND
The spruce is a forest tree that covers large areas of the European mainland. The Cep (small picture) is very common here, as is the Bay Bolete, the Russet Brittle Gill, the False Saffron Milk Cap and the highly poisonous Death Cap.

BIRCH WOODLAND
In addition to the well-known Brown Birch Bolete (small picture), which grows around individual birches or in sparse birch woodland, this is also a favoured habitat for the Fly Agaric, Woolly Milk Cap and other Milk Caps.

BEECH WOODLAND

Either alone or with other deciduous trees, beeches form a useful habitat which should be protected. Lurid Bolete (small picture), Summer Bolete, Charcoal Burner and Pale Milk Cap are just some of the fungi that commonly grow around beeches.

PINE WOODLAND

Many edible species of mushroom grow in pinewoods, including Slippery Jack (small picture) and related species, such as the Granulated Bolete, Shallow-pored Bolete and the Saffron Milk Cap.

similar, but deadly, Funeral Bell mainly grows on dead spruce.

A variety of habitats

Fungi are not only found in woodland, they inhabit almost every other type of habitat. Whereas some prefer open grassland, whether fertilised or unfertilised, other species, such as the tasty Field Mushrooms, grow in fields and on moorland. Some can live in salt marshes and thrive in dunes near the coast, while others, such as the Oyster Mushroom, can survive the cold and even sprout during the frost.

Fungi can grow on doorsteps, in gardens, parks or at the roadside. These are all favourite habitats for the Common Morel and Shaggy Ink Caps.

A threatened existence

Encroaching urbanisation, restrictions on the growth of species of tree that are not seen as important for the economy, increased planting of pure spruce woodland, the increased fertilisation of previously unfertilised grassland and air pollution are all factors that have caused many species of fungus to become endangered. The reduction in species numbers has n○ been halted, despite attempt do so by restricting or ba○ mushroom-picking. In ○ has little effect, since ○ green plants and the○ actually spread by

> Fly Agaric contains ibotenic acid and traces of muscimol

Poisonous fungi and fungal poisons

There are only about a dozen deadly poisonous fungal species in Europe, but before you begin to collect mushrooms, you should make sure that you know which ones are dangerous.

Beautiful but deadly

Cases of mushroom poisoning are usually the result of insufficient knowledge and the complacency of mushroom pickers. Many people judge mushrooms mainly on their appearance. If a mushroom ~~ks~~ and smells good, some ~~le~~ think that it is edible. ~~unately~~, there is no relation- ~~ween~~ the appearance and ~~~ mushroom, and ~~~poisonous or not. or other animals without suffering any ill effects is also irrelevant. The same mushroom may well be poisonous for humans.

The most important fungal poisons

The Death Cap mushroom contains the poison amanitine, as does the Funeral Bell. This poison can withstand heat and loses none of its toxicity even when dried for several years. The symptoms of poisoning begin 12 to 15 hours after consumption. Death is caused by liver failure.

Gyromitrin, a poison found in the False Morel, was only identified about 20 years ago. If consumed in large quantities it can be deadly.

Muscimol was first identified in Fly Agaric, which only contains traces of the poison. Many

Inocybes contain higher levels of muscimol and are much more dangerous. Muscimol quickly affects the circulation and, in high doses, can cause death by heart failure. Fly Agaric and Panther Cap also contain ibotenic acid, which mainly affects the central nervous system. It produces hallucinations, though poisoning is rarely fatal.

Orellanine is the name given to a group of fungal poisons found in *Cortinarius orellanus* and related species. They can cause lasting kidney damage. Symptoms of poisoning may not appear for one or two weeks, but are often fatal.

Consume at your risk

Some mushrooms – including the Devil's Bolete, as well as certain allegedly edible mushrooms – can cause serious stomach upsets, or even poisoning, if eaten raw. So never eat mushrooms raw while out collecting. Otherwise edible mushrooms may become dangerous if they are picked when they are starting to decompose. Only pick young and fresh mushrooms, in reasonably dry conditions, and cook them for at least 15 minutes. In recent years, there has been much talk of environmental toxins and radiation being absorbed by fungi. Species of fungi in some areas may contain high levels of cadmium, lead, mercury and other heavy metals, or the radioactive caesium. Never pick fungi for eating purposes if they are growing beside a busy road.

> After eating the Common Ink Cap, alcohol should not be consumed for two days, or you risk being poisoned

> Brown Roll-rim can cause life-threatening poisoning if eaten in large quantities

WHAT TO DO IF YOU ARE POISONED BY A MUSHROOM

- Empty the stomach immediately by vomiting.
- Go to your doctor or nearest hospital immediately, and explain the problem.
- Keep part of the mushroom or stomach contents, and take them with you to the hospital.

The dangers of misidentification

Destroying Angel (p62)

Field Mushroom (p48)

Death Cap (p62)

Charcoal Burner (p132)

Funeral Bell (p54)

Two-toned Pholiota (p54)

☠ 🍴

False Morel (p176)

Common Morel (p178)

Panther cap (p66)

The Blusher (p64)

Deadly Fibrecap (p70)

St George's Mushroom (p70)

List of fungi

Cep
Porcini | Penny Bun Mushroom
Boletus edulis

APPEARANCE: Cap is hemispherical in young specimens ①, but broadly convex when mature; pale brown to chestnut and about 20cm in diameter; surface sometimes slightly slimy; tubes have narrow pores, initially white or whitish-grey, but later yellow or orange and green ④; does not darken when pressed. Stem is normally very bulbous, as wide as the cap, though occasionally club-shaped.

FLESH: White, with a nutty smell and taste.

DISTRIBUTION: Common in coniferous woods, mainly under spruce and pine.

TYPICAL FEATURES
The upper third of the stem of the Cep is covered in a white reticulated pattern ③, which can be found in the centre of the stem in older specimens.

EDIBILITY: The Cep is one of the tastiest and highest valued edible species.

SIMILAR SPECIES: Pine Bolete (*Boletus pinophilus*) ② has a dark, red-brown stem with a very faint reticulation. It grows mainly under pine and fir trees.

Bitter Bolete
Tylopilus felleus

APPEARANCE: Cap of young mushroom is hemispherical, but later flattens. Up to 10cm in diameter ①; beige to reddish-brown with a smooth, felt-like surface, which is rarely slimy, even when wet. Tubes are initially a dirty white, but turn a pinkish-brown when mature ④. Stem is narrow, with extensive, dark reticulation. ③.

FLESH: Whitish ②; smell is pleasant, but the taste is extremely bitter.

DISTRIBUTION: Often seen growing in straw nd moss, especially under pines and spruces.

TYPICAL FEATURES
Bitter Bolete has pink tubes when fully grown. Placing the tip of the tongue on the skin of the cap is sufficient to establish the bitter taste.

IBILITY: Although the Bitter Bolete contains small traces of muscimol, Fly Agaric poison, it is not really poisonous. It is unpleasantly bitter and a small piece of the mushroom can make a whole meal inedible.

R SPECIES: For non-experts, the Bitter Bolete is easily confused ous species of Cep. Tasting a small piece can prevent confusion.

Bitter Beech Bolete
Boletus calopus

APPEARANCE: Cap is pale grey to greyish-ochre, initially hemispherical but flattening later and broadly convex, with a diameter of up to 15cm. The skin of the cap is dry, silky, and matte. Tubes are pale yellow in young specimens but later turn olive-yellow or olive-green; blue-green flecks appear when it is pressed ②. Stem is bulbous to cylindrical; pale yellow near top with a red reticulation ③. The base of the stem is usually bright red ①.
FLESH: Yellowish-white, turning slightly blue when cut ④; no aroma and a very bitter taste.
DISTRIBUTION: Mainly in low mountain ranges and mountainous regions, in coniferous and deciduous woods, mostly under spruces, firs and copper beeches; rare in lowland areas.
EDIBILITY: The Bitter Beech Bolete is inedible, and slightly poisonous when raw; it can cause serious indigestion.

TYPICAL FEATURES
The best way to distinguish the Bitter Beech Bolete from the very similar Devil's Bolete is by its yellow pores.

Devil's Bolete
Satan's Boletus | *Boletus satanas*

APPEARANCE: Cap is hemispherical or thick and flat, and up to 25cm wide; pale grey to greyish yellow ①, turning pale brown with age; surface is matte and dry. Tubes are yellow, usually with bright orange-red or carmine-red pores that turn blue when pressed. Stem is very thick, bulbous and stocky; initially yellow but later turns carmine red in the centre; surface has yellow reticulation ③.
FLESH: Yellow-white; slowly turns blue when cut ②; young flowers smell of urine, older plants smell of rancid fat.
DISTRIBUTION: Devil's Bolete is rare; it grows in warm, deciduous woodland, mainly under copper beeches and oaks.
EDIBILITY: This eye-catching mushroom is poisonous and can cause very acute stomach upsets, particularly if eaten raw.
SIMILAR SPECIES: False Satan's Bolete (*Boletus splendidus*) ④ has a mouse-grey cap when young, which turns red from the margins inwards.

TYPICAL FEATURES
Devil's Bolete grows in deciduous woods. It has a white cap, red pores and a stem that is carmine red in the middle. It is protected.

Scarlatina Bolete ⅋⅋

Boletus erythropus (B. luridiformis)

APPEARANCE: Cap initially hemispherical, but later becoming broad and flat ① and up to 10cm in diameter; grey-brown to dark brown in colour; silky when dry, only slightly slimy when it rains. Tubes are yellow with bright red pores; immediately turns dark blue when touched. Stem is cylindrical or club-shaped, yellow and covered in fine, red flakes ③.

TYPICAL FEATURES
The stem is floccose, there is never any reticulation.

FLESH: Yellow when first cut but turning dark blue within seconds ④; after a while, the blue colour disappears; smell and taste are mild and pleasant.

DISTRIBUTION: In deciduous and coniferous woodland; common on high ground.

EDIBILITY: The blue colour may act as a deterrent, but this species is a delicious edible mushroom when cooked.

SIMILAR SPECIES: The stem of the rare Deceiving Bolete *(Boletus queletii)* ② does not have reticulation or flakes on the surface.

Lurid Bolete ⅋⅋

Boletus luridus

APPEARANCE: Cap up to 20cm wide, hemi-spherical or broadly convex and bulbous; slightly paler than that of the Scarlatina Boletus and olive-brown; dry and silky. Tubes are yellow to olive-green ②, with orange-red pores, olive-coloured on older specimens. Stem is thick and bulbous to club-shaped ①, yellow in colour with an elongated, dark red-brown reticulation ③.

TYPICAL FEATURES
Lurid Bolete can be identified by its prominent red-brown reticulation on the stem ③ and by a maroon line about the tube layer.

FLESH: Flesh has an insipid taste and scent; yellow, but turns blue when cut.

DISTRIBUTION: Relatively common in deciduous and coniferous woods; often found in gardens and parks.

EDIBILITY: Mildly poisonous when raw, but when thoroughly cooked it is considered by many to be an excellent edible mushroom. Can cause stomach upsets in some people.

SIMILAR SPECIES: Pink-capped Bolete *(Boletus rhodoxanthus)* ④ is different only due to its pink-coloured cap.

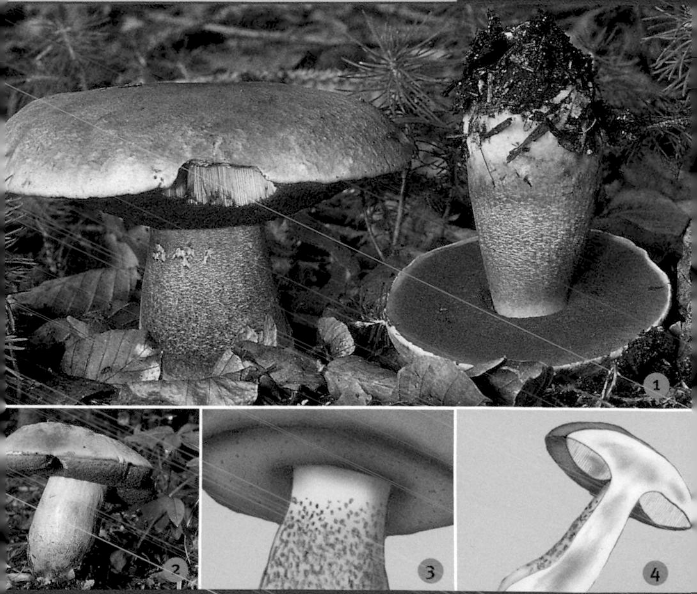
1
2
3
4

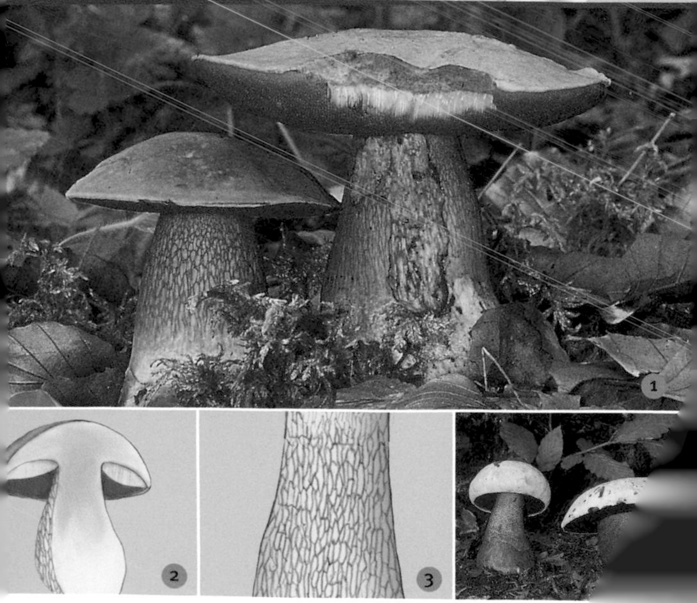
1
2
3

Summer Bolete

Boletus reticulatus

APPEARANCE: Cap is pale to medium brown ① and initially hemispherical, but later broadly convex in older specimens; only rarely larger than 15–20cm; fine matte surface that can flake off in patches when dry ③. Tubes sinuate where they join the stem; grey or grey-white in young fruiting bodies, but yellowy or a pale olive-green in older specimens ④. Stem is pale brown, club-shaped or bulbous and covered in elongated reticulation ②.

TYPICAL FEATURES
Unlike the Cep (p26), the Summer Bolete grows mainly under oaks and copper beeches.

FLESH: Firm and white or cream-coloured; slightly brown beneath the skin of the cap; smell and taste are pleasant and mild.

DISTRIBUTION: In deciduous woodland, mainly under oaks and copper beeches. Uncommon but widespread.

EDIBILITY: An excellent and delicious edible mushroom, suitable for many different uses and for drying.

Bay Bolete

Xerocomus badius

APPEARANCE: Cap 5–10cm wide, very occasionally up to 15cm; hemispherical and dark brown ①; normally silky and dry, but becomes greasy or slimy in wet weather. Tubes are very compact, pale yellow or whitish in young specimens, later turning yellow, olive-yellow or a pale olive-green. Stem is brown, normally cylindrical, but occasionally slightly bulbous; surface is grainy or slightly floccose ③.

TYPICAL FEATURES
The yellow tubes of the Bay Bolete turn bluish-green in a few seconds when pressed.

FLESH: The white flesh ② turns pale blue when sliced. Smell and taste are mild and pleasant.

DISTRIBUTION: Common in coniferous woodland, mainly in upland forests, under pines and spruces; also found beneath beeches and oaks.

EDIBILITY: This excellent edible mushroom is suitable for all culinary uses and for drying.

SIMILAR SPECIES: The inedible Dusky Bolete (*Porphyrellus porphyrus*) ④ can be recognised by the grey-brown colour of the whole fruit body, including the tubes.

Red Cracking Bolete
Xerocomus chrysenteron ⅋⅋

APPEARANCE: Cap is 3–8cm wide, but can vary greatly in colour, from yellow-brown to dark olive to almost black; surface is dry and matte, and always cracked ④. The cracks eventually turn red. The tubes are compact and initially pale yellow, but later olive-green; the cap turns blue when pressed. Stem is normally cylindrical, but very occasionally broader at the base; base colour is yellow but always has carmine-red patches ①.

FLESH: The very soft flesh turns slightly blue when cut ③. The taste is faintly acidic but there is hardly any smell.

DISTRIBUTION: In deciduous and coniferous woodland.

EDIBILITY: Younger specimens are best for cooking. Good for drying.

SIMILAR SPECIES: The edible Matte Bolete *(Boletellus pruinatus)* ② has a dark brown, frosted cap and normally has a red stem.

TYPICAL FEATURES
The carmine-red stem and the red cracks in the skin of the cap are distinguishing features of this bolete.

Suede Bolete
Xerocomus subtomentosus ⅋⅋

APPEARANCE: Cap is a light olive-brown, occasionally also chestnut brown; 5–10cm wide ①. Tubes are chrome yellow, with conspicuously large, angular pores ④, which do not normally change colour when pressed. Stem is cylindrical, full and fleshy and rarely straight. It is pale brown in colour and has vertical ridges, mainly near the top ②.

FLESH: The flesh has a pleasantly mild smell and taste. It is white or pale yellow in colour ③ and slightly firmer than that of the Red Cracking Bolete. It only turns a slightly pale blue-green colour. The flesh of the stem is very yellow.

TYPICAL FEATURES
The cap surface of the Suede Bolete is always dry and felt-like. It is never slimy and does not peel off in patches like the Red-cracking Bolete.

DISTRIBUTION: Suede Bolete is widespread in coniferous and deciduous woodland, from sea level to mountain regions. Normally found in association with bilberries.

EDIBILITY: This pleasant edible mushroom has a full-bodied flavour and is suitable for any kind of preparation, including drying.

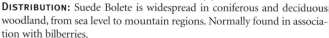

Brown Birch Bolete
Leccinum scabrum ᵱᵞᵱ

APPEARANCE: Cap is up to 12cm in diameter, in a variety of shades of grey or brown ①. The surface is smooth, matte and dry, but often slimy in older specimens. Tubes are initially whitish-grey, turning greyish-brown with age, and sinuate around the stem. In older specimens, the tubes swell and protrude from under the cap ③. Stem covered in small white or grey scales ④.

TYPICAL FEATURES
This fungus only grows under birches and has a relatively long, narrow stem, which is densely covered in scales.

FLESH: The white flesh has a very faint smell and a mild taste. It often turns blue-green at the base of the stem.

DISTRIBUTION: Grows exclusively under various species of birch; sometimes also found in grassland.

EDIBILITY: Brown Birch Bolete is a delicious edible mushroom.

SIMILAR SPECIES: Hornbeam or Grey Bolete *(Leccinum griseum)* ② is also edible and turns grey-violet or almost black when cut.

Orange Birch Bolete
Leccinum versipelle ᵱᵞᵱ

APPEARANCE: Cap is yellow-orange to brick-red ① and hemispherical in young specimens, later becoming wide and bulbous and up to 20cm wide; skin extends beyond the tubes at the margin ③; surface is dry and matte, only slightly sticky after rain. Tubes are white, the pores later turning ochreous-grey. Stem is white, with blackish-brown floccose scales ①.

TYPICAL FEATURES
Orange Birch Bolete always grows under birches and has a stem covered in prominent black scales. The base turns blue-green when cut.

FLESH: The flesh has a pleasant taste and smell and is whitish in colour. It turns blue-green when cut ②.

DISTRIBUTION: Found exclusively under birches, often with Brown Birch Bolete.

EDIBILITY: Orange Birch Bolete has a very pleasant flavour, although the flesh turns an unappetising greyish-black when cooked.

SIMILAR SPECIES: Orange Oak Bolete *(Leccinum quercinum)* ④ grows in eciduous woodland, mainly under oaks. The flesh turns red when cut.

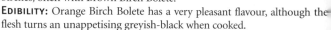

Bovine Bolete
Suillus bovinus

APPEARANCE: Cap is a coppery reddish-brown colour; hemispherical when young, later becoming broad and bulbous and reaching a diameter of 5–8cm; slightly slimy when damp. Tubes are greyish-yellow, turning copper brown with age; very large pores ③ with uneven, protruding edges, slightly decurrent at the stem. Stem is narrow and cylindrical and the same colour as the cap ①.
FLESH: The flesh has a pleasant taste and smell. It is pale yellow, turning reddish in places ④. Flesh also turns slightly red when cut.

TYPICAL FEATURES
Bovine Bolete only grows under pines and can be identified by its large pores with uneven, protruding edges.

DISTRIBUTION: Only under pines, often on sandy and marshy soils.
EDIBILITY: Edible, but only has an average flavour and is mainly recommended for mixing with other species. Slightly slimy when cooked.
SIMILAR SPECIES: The Sand Bolete *(Suillus variegatus)* ② has a slightly grainy, matte, sand-coloured cap and a dark olive stem.

Weeping Bolete
Suillus granulatus

APPEARANCE: Cap is initially hemispherical, later becoming flatter and broader ①; 4–10cm wide; yellow-brown to rust-brown and slightly greasy when damp. Tubes are pale yellow on young specimens and often produce milky droplets ④; tubes later turn darker olivaceous-yellow. Stem is pale yellow and covered in tiny dark spots near the top ②.
FLESH: The flesh is yellowish-white, yellower towards the stem ③. Flesh of young fruiting bodies is very soft, smelling and tasting pleasantly mild and slightly fruity.

TYPICAL FEATURES
Weeping Bolete can be easily distinguished from Slippery Jack (p40), which also grows under pines, because it does not have a ring on the stem.

DISTRIBUTION: Weeping Bolete almost always grows under pines. It prefers limestone soils and is often found together with Slippery Jack (p40).
EDIBILITY: The flesh is very soft and has little flavour and so the young fruiting bodies are recommended for mixing with other species. Eaten in large quantities, the mushroom can have a laxative effect.

Larch Bolete
Suillus grevillei

APPEARANCE: Cap is initially hemispherical and convex, later becoming broader and bulbous ①, and up to 12cm wide; pale orange-yellow to golden brown; greasy in damp weather. Tubes have compact pores and are golden to olive-yellow ②. Cylindrical stem is similar in colour to the cap, with a thin white, fugacious ring ③ that disappears in older specimens.

TYPICAL FEATURES
This fungus grows under larches and can be identified by the gold-coloured surface of the cap and the similarly coloured tubes.

FLESH: Flesh is pale yellow, especially in the stem ④; smell and flavour mild and pleasant; does not discolour when pressed. Flesh is very soft on older specimens.

DISTRIBUTION: Larch Bolete grows on high and low ground and is only found under larches.

EDIBILITY: The flesh is very soft and has little flavour, so the young fruiting bodies are recommended for mixing with other species. The skin of the cap can be removed before cooking.

Slippery Jack
Suillus luteus

APPEARANCE: Cap is hemispherical, chocolate or orange-brown ①, and up to 10cm wide; very greasy in damp weather. In young specimens remnants of the veil often cling to the cap ③. Tubes are yellow ④, olive-yellow on older mushrooms, with very narrow pores. Stem is normally cylindrical, yellow near the top, with dark spotting; below the prominent ring, the stem has a slimy coating.

TYPICAL FEATURES
Slippery Jack is only found under pines and can be recognised by its dark brown, greasy cap, and prominent ring on the stem.

FLESH: The whitish-yellow flesh ② does not discolour when cut. The smell and taste are pleasant and mild.

DISTRIBUTION: Exclusively under pine trees.

EDIBILITY: Young specimens are excellent for cooking, but older mushrooms are very soft. The skin of the cap should be removed before cooking. In recent years there have been reports that repeated consumption of large quantities of Slippery Jack can cause allergic reactions.

Hollow Bolete

Boletinus cavipes

APPEARANCE: Cap is conical in young specimens, but broader and bulbous in older mushrooms ①; 4–10cm wide; quite variable in colour, from chestnut brown to orange or sulphur-yellow; surface is silky or matte and dry. Tubes are initially yellow, later turning olive-green, with very wide angular pores and ridges running from stem to margin ②. Stem is hollow ④, with a whitish ring just below the tubes ③; below the ring, the stem is the same colour as the cap.

FLESH: The flesh has very little or no smell but a pleasant taste. It has a pale yellow colour.

TYPICAL FEATURES
Hollow Bolete is hard to mistake, due to its hollow stem, dry, matte cap and the ring at the top of the stem.

DISTRIBUTION: Found on high and low ground, always in association with larches.

EDIBILITY: This mushroom has a good flavour but a low yield. It is not recommended for drying.

Alder Bolete

Gyrodon lividus

APPEARANCE: Cap 3–12cm wide; margin often inrolled on young specimens, but is later bent and angled ①; yellow to ochre-brown, turning brown when pressed; surface is matte and dry, occasionally greasy in old specimens. Tubes short and compact, very decurrent at the stem; golden yellow ④, later olive-coloured, and immediately turning blue-black when pressed. Stem is the same colour as the cap, and often wrinkled.

TYPICAL FEATURES
The tubes of the Alder Bolete are hard to separate from the cap flesh. The cap surface is often pitted ③.

FLESH: Yellow in colour, turning bright blue when cut ②; inconspicuous aroma and fairly bitter flavour.

DISTRIBUTION: Relatively rare; forms a symbiotic relationship with alders and grows in damp, marshy habitats, but also in forests.

EDIBILITY: This mushroom has long been considered edible but of poor quality. It contains certain substances very similar to those in the Brown Roll-rim (p94), which can cause life-threatening damage if consumed too often. Experts warn strongly against eating this mushroom.

Sheep's Bolete

Albatrellus ovinus

APPEARANCE: Cap 4–10cm wide; young specimens are greyish-white, smooth and bulbous, yellowing with age and irregularly shaped; cap surface is cracked. Tubes are very short with narrow pores, decurrent at the stem. Stem is white, broadening where it joins the cap ①.

FLESH: White in younger specimens, later turning lemon-yellow, mainly in the stem ④. Smells of almonds and tastes pleasantly of mushroom.

TYPICAL FEATURES
Only grows in conifer woodland and has a flesh which turns yellow and which does not taste bitter.

DISTRIBUTION: Sheep's Bolete is commonly found in conifer woodland in mountain regions. Often forms fairy rings. Not found in the UK.

EDIBILITY: The mushroom is not bitter, has a pleasant taste, and is used in cooking, but it can cause serious stomach upsets in some people.

SIMILAR SPECIES: Crested White Bolete *(Albatrellus cristatus)* ② grows in deciduous woodland and has an olive-brown cap, which becomes cracked with age ③.

Fringed Polypore

Polyporus lepideus

APPEARANCE: Cap has thin flesh; grey-brown to ochre-brown ① and only 2–6cm in diameter; surface is smooth and matte, sometimes with bristle-covered wrinkles at the margin. Tubes are very short and white ②, with pores so narrow they can only be seen through a magnifying glass; tubes slightly decurrent at the stem. Stem is full-fleshed and brown, with a smooth, matte or floccose surface ③.

TYPICAL FEATURES
The most distinguishing feature of Fringed Polypore is its white, narrow pores ②.

FLESH: Whitish with only a faint aroma; has a tough or elastic texture.

DISTRIBUTION: Fringed Polypore is normally saprophytic and is common on dead or rotten deciduous wood ①; mainly on beech trunks.

EDIBILITY: The fruiting bodies of the Fringed Polypore have a very tough flesh and are therefore inedible.

SIMILAR SPECIES: Winter Polypore *(Polyporus brumalis)* ④ has larger pores, which can be seen with the naked eye. The cap is hairless and darker than the Fringed Polypore. Grows from October to late March.

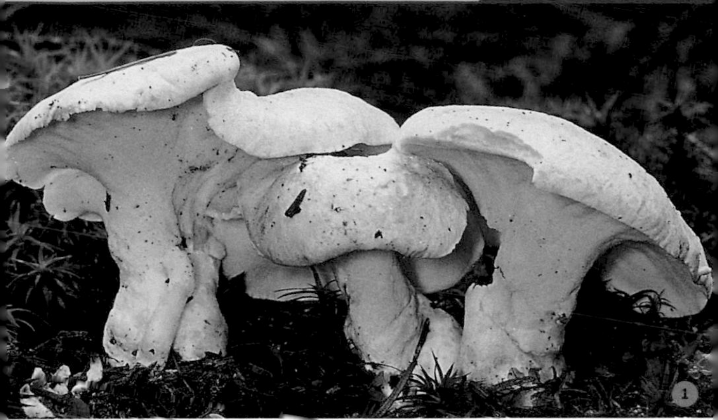
1

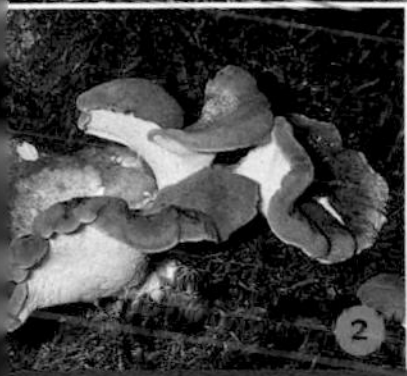
2

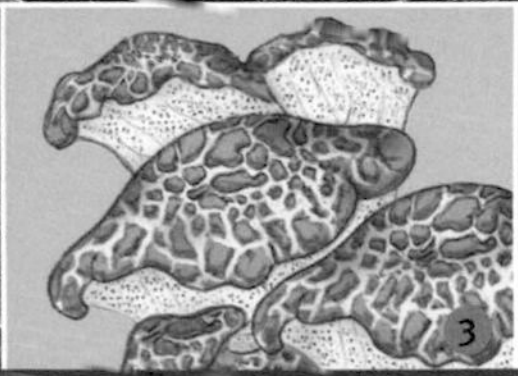
3

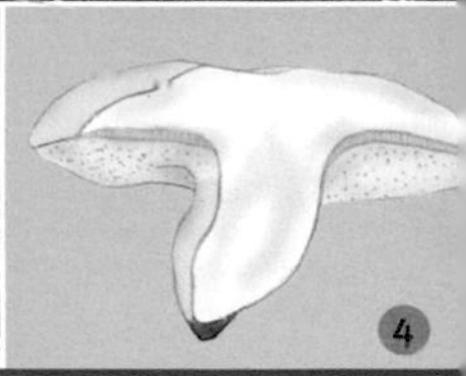
4

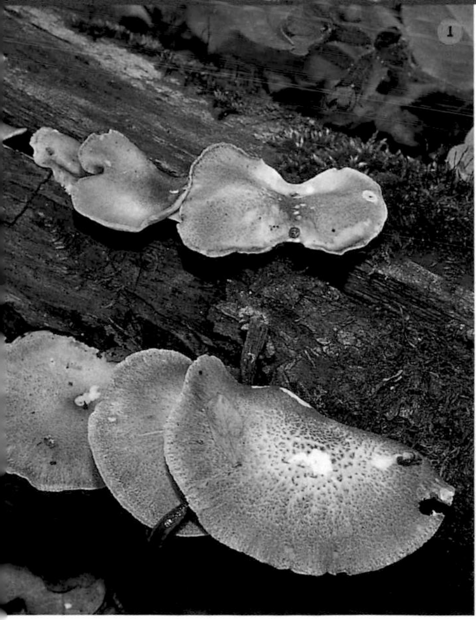
1

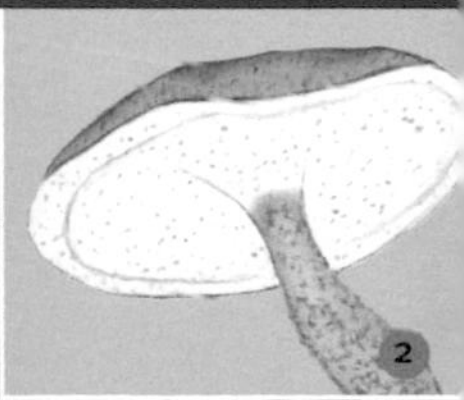
2

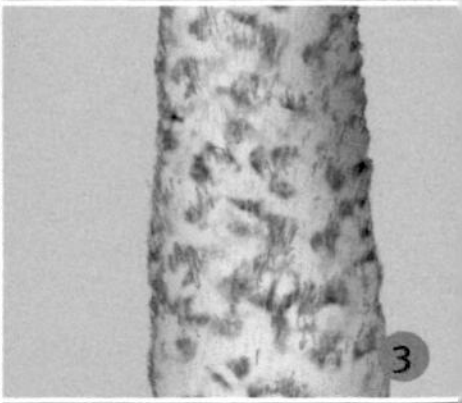
3

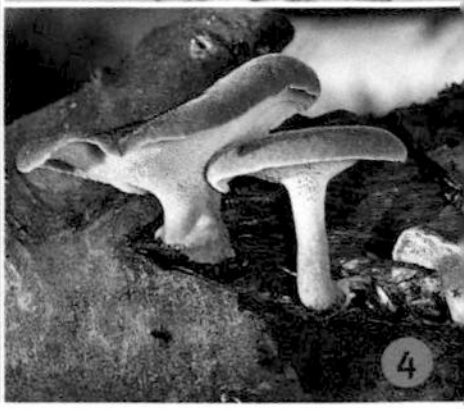
4

Wood Mushroom
Agaricus essetii

APPEARANCE: Cap is white and is initially almost spherical ①, but later becomes flatter and broader, often slightly yellowing at the margin ④; just over 10cm in diameter; surface is smooth and silky. Gills are compact, and pale grey or light pinkish-brown in younger specimens ②, turning dark brown or almost black. Stem is cylindrical or slightly wider at the base; has an obvious, hanging ring ① and normally an angular base ③.

TYPICAL FEATURES

The cap of the Wood Mushroom turns a bright chrome yellow colour when pressed.

FLESH: White, does not discolour; pleasant smell of aniseed.

DISTRIBUTION: Widespread in coniferous and mixed woods.

EDIBILITY: This species was long considered a good edible mushroom. However, laboratory tests have shown that the substance which causes the skin of the cap to turn yellow is highly carcinogenic. Consumption of the mushroom is therefore not recommended. It also has relatively high heavy metal (cadmium) contamination.

Yellow-staining Mushroom
Agaricus xanthoderma

APPEARANCE: Cap is white with a smooth surface; younger specimens are convex, and often flat on top ①; with age they become flatter and broader and up to 12 cm wide; turns yellow when pressed. Gills are compact, initially pink but later brownish-black. Stem is narrow with a prominent, white, double ring; often has a bulbous base.

TYPICAL FEATURES

The yellow colour at the base of the stem ② distinguishes this fungus from the edible Field Mushroom. It can also be identified by the smell of ink or carbolic acid which is even more noticeable when the fungus is cooked.

FLESH: The flesh is white and smells of carbolic acid or ink. The base of the stem turns bright yellow when cut ②.

DISTRIBUTION: Saprophytic and found in meadows, heathland and in parks and gardens.

EDIBILITY: Poisonous and causes vomiting shortly after consumption.

SIMILAR SPECIES: The pleasant-tasting Horse Mushroom (*Agaricus arvensis*) ④ smells of aniseed. It also has a bulbous base and the mushroom does not turn bright yellow when damaged ③.

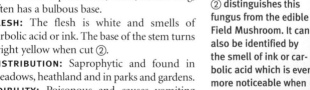

Field Mushroom 🍴🍴

Agaricus campestris

APPEARANCE: Cap is almost spherical in younger specimens, but later broadens ①, and is up to 10cm in diameter; pure white and does not discolour. Gills are very compact and pinkish-grey ④, later turning dark brown to almost black. Stem does not have bulbous base ③, but is often slightly curved, with a white, membranous, hanging ring.

FLESH: The white flesh has a pleasant odour and turns slightly reddish when cut.

DISTRIBUTION: Grows in meadows and heathland, often in large numbers; mainly found after rainfall at the end of a dry spell.

TYPICAL FEATURES

The Field Mushroom is easy to identify because it grows in meadows. The flesh has a pleasant smell and turns slightly reddish when cut.

EDIBILITY: An extremely tasty edible mushroom with a very high yield.

SIMILAR SPECIES: Macro Mushroom (*Agaricus macrosporus*) ② grows slightly larger and can be identified by its scaly cap, thick stem and white flesh that turns creamy when cut.

Blushing Wood Mushroom 🍴🍴

Agaricus silvaticus

APPEARANCE: Cap is rarely more than 10cm in diameter; in young specimens it is convex and inrolled ②, but later becomes broadly convex to flat ①; brown with dark, floccose but well-anchored scales on the surface. Gills are compact and do not join the stem ④; initially very pale brown to grey, but later turning chocolate brown. Stem is white and sometimes slightly scaly or covered in woolly flakes; up to 12cm tall, cylindrical and hollow ③, with a slightly bulbous, thicker base and a wide hanging ring.

TYPICAL FEATURES

The flesh of all parts of the Blushing Wood Mushroom turns a uniform dark red when cut.

FLESH: The pure white flesh has a mild smell and taste. It slowly turns blood-red or maroon when cut.

DISTRIBUTION: This common mushroom grows in groups in coniferous woodland on limestone soils. It can occasionally be found under beeches.

EDIBILITY: Blushing Wood Mushroom has a high yield and is a delicious edible mushroom.

Spring Fieldcap

Agrocybe praecox

APPEARANCE: Cap is initially broadly convex, but later becomes flatter; scarcely 7cm wide; ochre-yellow to ochre-brown, often with an umbo in the centre; surface is dry and smooth. Gills are very compact, pale clay colour, and dirty-brown or blackish-brown in older specimens ①. Stem is whitish, and younger specimens have a core near the top, but older mushrooms have a completely hollow stem.

TYPICAL FEATURES
Older specimens often only retain the ragged remnants of the thin, ribbed ring ②.

FLESH: Often has a strong or quite strong smell of flour; often unpleasantly bitter.

DISTRIBUTION: Mainly found in light deciduous woodland, in gardens and parks, on meadows and heathland; grows in clusters.

EDIBILITY: Often bitter. Edible but not particularly tasty.

SIMILAR SPECIES: Bearded Field Cap *(Agrocybe dura)* ④ is similar, but does not have a floury smell, and the ring zone is barely visible. The cap often has scaly cracks ③.

Rough-ringed Agaric

Stropharia rugosoannulata

APPEARANCE: Cap is initially broadly convex, but later becomes broader and conical; 5–15cm wide; grey-brown to reddish-brown ①, very occasionally yellow-brown ③; shiny, and mostly dry to the touch, but can be greasy in wet weather. Adnate gills are pale greyish-violet in colour ④, becoming grey-black with age. Stem is thick and whitish near the top, but increasingly yellow near the base; up to 15cm tall and 2.5cm wide.

TYPICAL FEATURES
The wide ring is initially white but later turns yellow ② and eventually greyish-black once the spores have been dispersed ①.

FLESH: The white flesh does not discolour when cut or when cooked; it has almost no smell and tastes slightly of soil, reminiscent of raw potatoes.

DISTRIBUTION: In the wild, the Rough-ringed Agaric occasionally grows in freshly harvested fields, but it is rare everywhere. It can be cultivated on bales of straw.

EDIBILITY: This edible mushroom has a high yield.

51

Shaggy Inkcap | Lawyer's Wig 🍴
Coprinus comatus

APPEARANCE: Cap is typically acorn-shaped or cylindrical and covered in floccose scales ①; about 10cm tall; young fruiting bodies are white, turning brownish at the top; cap becomes bell-shaped with age, cracking at the margin ③. Gills are compact and initially white, but later turn pink ②; older specimens turn black from the margin inwards and the flesh melts ④. Stem is hollow, with vertical striping and a low ring.

TYPICAL FEATURES
The ring is positioned low down on the tubular stem of young fruitbodies and is very thin. It can fall off very early.

FLESH: The white flesh has a pleasant and mild smell and taste.

DISTRIBUTION: On meadows and heathland, beside footpaths and in gardens. Widespread species, often found in large numbers.

EDIBILITY: Young fruiting bodies are delicious edible mushrooms. When the gills begin to turn black and melt, the mushroom begins to disintegrate rapidly and is no longer edible.

Honey Fungus ◯
Armillaria mellea

APPEARANCE: Cap is initially convex, but later becomes broader; honey-yellow to brown, with fine, dark scales. Gills are creamy-white to flesh-coloured and adnate. There is a ring on the stem. Several stems can emerge from one point, or grow in clusters ①.

TYPICAL FEATURES
The scales on the cap can be rubbed off easily. The white ring is woolly or cottony ②.

FLESH: The brownish flesh has a faint smell, and the taste is unpleasant if chewed for a long time. The stem is tough and is not eaten.

DISTRIBUTION: The fungus is parasitic or saprophytic on dead deciduous and coniferous wood and it can kill a tree. Often found in very large numbers.

EDIBILITY: The Honey Fungus is highly prized in Spain as a good edible mushroom with a high yield. It is, however, mildly poisonous when eaten raw, and can even cause indigestion when cooked.

SIMILAR SPECIES: Shaggy Scalycap (*Pholiota squarrosa*) ④ is not edible and has a scaly, ringless stem, yellow flesh ③ and greyish-brown gills. Mainly grows on conifers.

Funeral Bell

Galerina marginata

APPEARANCE: Cap is initially convex, but later becomes broader and flatter. The flesh is thin ①; cap is bald and smooth; honey-brown when wet, but a pale yellow-brown when dry; rarely wider than 3cm. Gills are initially pale brown ②, but turn rust brown with age, and adnate at the stem. The ring falls off easily ④; above the ring, the stem is brownish and smooth; below it is dark brown, with a flaky or fluffy surface; hollow ③.

FLESH: The ochre-yellow or brown flesh smells strongly of flour.

DISTRIBUTION: Almost always on dead conifers, occasionally on deciduous trees.

EDIBILITY: The Funeral Bell contains amanitine, the poison in the Death Cap mushroom, and can cause fatal cases of poisoning.

TYPICAL FEATURES

The Funeral Bell can form dense carpets, but the stems always form individually. The floury smell is a further feature that helps distinguish this poisonous mushroom from the Two-toned Pholiota.

Two-toned Pholiota

Kuehneromyces mutabilis

APPEARANCE: Cap is broadly convex and 3–6cm wide; the margin on young specimens is covered in fine scales, but later becomes smooth; honey-brown when wet, turning pale at the top and yellow-brown at the margin ①. Gills are compact and cinnamon-coloured ③, adnate at the stem. Stem is narrow, and whitish-brown above the prominent ring; dark brown and scaly below the ring ②.

FLESH: The brown fibrous flesh of the stem has a pleasantly mild, floury smell and taste.

DISTRIBUTION: This common mushroom grows almost exclusively as a saprophytic fungus on dead deciduous wood, and is only rarely found on conifers.

TYPICAL FEATURES

Two-toned Pholiota can be distinguished from the dangerous Funeral Bell because it has a bushier growth, a more prominent ring, and does not smell as strongly of flour.

EDIBILITY: The caps of Two-toned Pholiota have a pleasant taste. They should only be collected by experienced pickers because they are very similar to the dangerous Funeral Bell and the two species can grow side by side ④ (see the darker mushroom on the right).

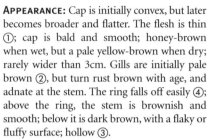

55

Golden Bootleg ⟨⟩

Phaeolepiota aurea

APPEARANCE: Cap is dry and has a fine, granulated texture; golden yellow in colour ①; younger specimens are convex and covered by a veil for a long time ②, later becoming umbonate ③, and up to 25cm wide. Gills are very compact, rusty yellow to pale rust-brown in colour and free of the stem. Stem is cylindrical, and slightly club-shaped at the base; can be more than 20cm high; surface below the yellow-brown ring has a grainy texture ④ and is the same colour as the cap.

TYPICAL FEATURES
The wide stem of the Golden Bootleg is grainy below the ring ④ and can be peeled away downwards.

FLESH: The whitish or yellowish flesh has a pleasant, slightly floury smell. It can be mild or have a strong mushroom flavour.

DISTRIBUTION: Relatively rare and saprophytic. It grows in clusters beside footpaths, and less often in deciduous and conifer woodland, in woodland clearings, in parks and gardens and in wood stores. Not found in Britain.

EDIBILITY: Edible and tasty, found mainly in central Europe.

The Gypsy ⟨⟩

Rozites caperatus

APPEARANCE: Cap is fleshy and a light honey-brown colour; up to 10cm wide; initially hemispherical ①, but quickly becoming broadly convex and almost flat; cracks radiate later from the margin inwards ②. Young fruiting bodies are covered in a pale lilac or silvery coating, which is easily washed off ④. Gills are compact, and have a slightly wrinkled appearance; they are the colour of milky coffee ③. Stem is cylindrical and up to 15cm tall; light brown, with a narrow, membranous ring.

TYPICAL FEATURES
Young fruiting bodies have a pale purple or silvery coating, which can be washed off ④. The Gypsy can be distinguished from the poisonous Gassy Web Cap (p114) by its ring.

FLESH: Flesh has pale brown marbling. Pleasantly mild smell and taste.

DISTRIBUTION: The Gypsy often grows in clusters, mainly in spruce and pine woodland and near bilberries; less common in deciduous woodland. Mainly found in Scotland.

EDIBILITY: This is one of the tastiest edible mushrooms and can be gathered in large quantities where it grows. Rare in southern England.

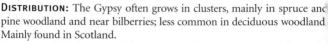

Saffron Parasol

Cystoderma amiantinum

APPEARANCE: Cap is yellow-ochre or orange-ochre ①; fine grainy texture; initially conical, becoming flatter and broader and 2–4 cm wide; centre is often umbonate ③. Gills white in young specimens, turning yellow with age; not compact and free of the stem. Stem is narrow and the same colour as the cap; golden-brown scales on the lower part of the stem.

TYPICAL FEATURES

The ochre-yellow to brownish stem has a faint ring. Above it, the stem is smooth, and below it, grainy and scaly ④.

FLESH: The yellowish flesh has an unpleasantly earthy smell and taste.

DISTRIBUTION: Found in coniferous woodland in autumn. Does not grow in the United Kingdom.

EDIBILITY: Saffron Parasol is inedible.

SIMILAR SPECIES: The Jason's Blistercap (*Cystoderma jasonis*) ② has pale, cream-coloured gills and can only be distinguished from the Saffron Parasol with a microscope. Mainly grows under spruces and pines.

Freckled Dapperling

Lepiota acutesquamosa

APPEARANCE: Cap is fleshy, and broad and flat in older specimens; 10–15cm in diameter; light brownish in colour, almost always with prominent dark rust-brown, conical or pointed scales ②; skin is often cracked, revealing the white cap flesh. Gills are very compact and often branching; white in colour ④, and free of the stem. Stem is up to 10cm tall and cylindrical; hollow and slightly bulbous at the base ③; the same colour as the cap, with a wide and membranous ring which

TYPICAL FEATURES

The mushroom has characteristic, pointed scales, and releases quite an unpleasant gassy smell.

is white on the upper side; ring never moves; stem is red-brown with fluffy scales beneath the ring ①.

FLESH: The soft, white flesh smells noticeably of gas. Do not taste.

DISTRIBUTION: This attractive mushroom grows in coniferous and deciduous woodland, but also in parks and gardens.

EDIBILITY: Freckled Dapperling is mildly poisonous.

59

Parasol Mushroom ¶¶¶
Macrolepiota procera

APPEARANCE: Cap whitish to pale brown and up to 25–30cm wide; initially convex, expanding to flat ①, with large scales and a dark umbo in the centre. Gills are whitish and very compact; free at the stem. Stem is white or brownish, and hollow; up to 40cm tall with grey-brown banding ④, and a double, removable ring ③.

TYPICAL FEATURES

The banded stem ④ and the moveable, double ring ③ are characteristic of the Parasol.

FLESH: Whitish; soft to brittle. Smell is pleasant with a nutty aroma; mild flavour. Stem is tough and fleshy and not suitable for consumption.

DISTRIBUTION: Relatively common; grows in sparse woodland, mainly in deciduous woods and on woodland edges; normally found in groups.

EDIBILITY: The Parasol is one of the best and most aromatic edible mushrooms. Should not be eaten raw.

SIMILAR SPECIES: Slender Parasol *(Macrolepiota mastoidea)* ② has a smooth stem and the cap has almost no scales.

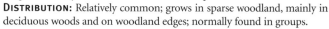

Shaggy Parasol ¶¶¶
Macrolepiota rachodes

APPEARANCE: Cap is up to 15cm wide, and is therefore smaller than the Parasol; very large scales ①; young specimens are convex and quite dark brown ②, becoming a lighter brown and broadly convex; older specimens are flat and broad. Gills are whitish and free at the stem; turn orange-red when pressed ④. Stem is narrow and hollow ③, up to 15cm high, and bulbous at the base; it has a double ring and is almost smooth; has no banding and is whitish to ochre in colour.

TYPICAL FEATURES

Unlike the Parasol, the Shaggy Parasol has no prominent hump in the centre of the cap, and the stem is not banded.

FLESH: The white flesh has a pleasant smell and delicious flavour. Slowly turns orange-red to saffron-brown when cut ④. The stem is very tough and should not be eaten.

DISTRIBUTION: Common in coniferous woodland; often in large numbers amongst the fallen needles.

EDIBILITY: Shaggy Parasol caps are not edible raw, but are delicious and perfectly digestible when cooked.

Death Cap

Amanita phalloides

APPEARANCE: Cap is olive-green or whitish green, often lighter at the margin ①; initially convex ②, but fully developed specimens are flat, broad and up to 15cm wide. Gills are compact, always pure white and free of the stem. Stem is thin and hollow ③, and banded like snakeskin ④, with smooth, hanging remnants of the universal veil (which sometimes fall off early); thick and bulbous volva, surrounded by a membranous veil.

TYPICAL FEATURES

The upper rim of the volva on the stem of the Death Cap is surrounded by a membranous sheath.

FLESH: In young mushrooms, the white flesh has almost no smell, but the fungus later smells slightly sweet like honey.

DISTRIBUTION: In deciduous woodland, mainly under oaks and beeches.

EDIBILITY: A single, small fruiting body is sufficient to kill a fully-grown human being. The fungus contains several toxins, the principal one being amanitine, which is not destroyed by cooking or drying.

Destroying Angel

Amanita virosa

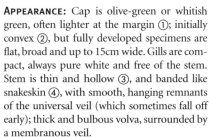

APPEARANCE: Cap is snow-white, and slightly yellow in the centre; young specimens are egg-shaped ①, and later conical ③ or broadly convex, but never completely flat; up to 10cm wide; skin of cap is often quite greasy. Gills are free, compact and pure white. Stem is also pure white, often with tattered remnants of the universal veil, below which the stem has woolly scales ②; broad and bulbous at the base, with a membranous white veil ④.

TYPICAL FEATURES

The gills of this poisonous mushroom are always pure white, and never grey, brown or pink like the gills of the Field Mushroom. The veil around the volva is also very characteristic.

FLESH: In young specimens, the white flesh has almost no smell, but it later smells slightly sweet like honey.

DISTRIBUTION: In coniferous woodland, usually on high ground.

EDIBILITY: The Destroying Angel and the Death Cap are the most dangerous of all fungi. Even tiny quantities can be deadly.

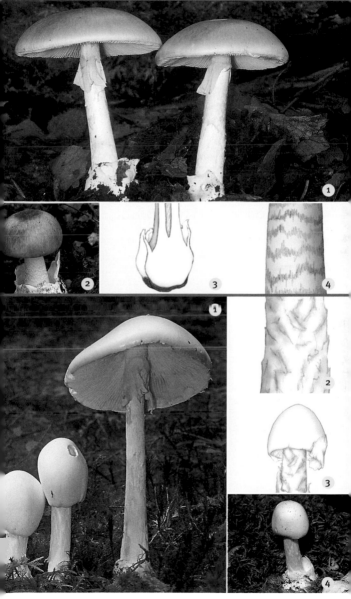

Fly Agaric
Amanita muscaria ☠

APPEARANCE: Cap is initially hemispherical, but later flattens and is up to 15cm in diameter; bright red with fluffy remnants of the veil ①; these remnants wash off in heavy rain. Gills are compact, pure white, and free of the stem. Stem has a ribbed, hanging ring; base broadens into a volva, which is fringed with the fluffy remnants of the veil ③.

TYPICAL FEATURES
If rain washes away the veil remnants, young Fly Agaric specimens can be safely distinguished from red Brittlegills (p128 – 136) by their ringed stems.

FLESH: White flesh has no smell, and is lemon yellow directly below the skin of the cap ②.
DISTRIBUTION: Coniferous and deciduous woods, under spruces and birches; common.
EDIBILITY: Fly Agaric is well-known to be poisonous, and the poison cannot be removed by cooking or by peeling the skin from the cap. The poisonous principle, muscarine, is strongly hallucinogenic but rarely fatal.
SIMILAR SPECIES: Caesar's Mushroom *(Amanita caesarea)* ④ has yellow flesh and gills and is an edible mushroom. It does not grow in Britain.

The Blusher
Amanita rubescens 🍴

APPEARANCE: Cap is hemispherical in young specimens, but flattens ①, and is 5–15cm wide; yellow-brown, fleshy-pink to dark reddish-brown in colour and covered in spots of reddish-grey veil remnants ②, which are easily washed off by the rain. The free gills are compact and are initially white, but later have reddish flecks. Stem is cylindrical, with a volva at the base ③; woolly below the ribbed ring ④.

TYPICAL FEATURES
The reddish colour of the Blusher, its broad, ribbed ring and the fact that the flesh slowly turns red enable it to be easily identified.

FLESH: Flesh has a pleasant, mild smell. Taste is initially sweet but becomes unpleasant. Whitish flesh slowly reddens when damaged.
DISTRIBUTION: Common in coniferous and deciduous woodland.
EDIBILITY: The Blusher is an edible mushroom, though experts differ as to how tasty it is. Because it is easily confused with the similar, but dangerously poisonous Panther Cap (p66), it should only be collected by experienced pickers.

Panther Cap

Amanita pantherina

APPEARANCE: Cap is greyish-brown to yellow-brown, with pure white veil remnants, which are easily washed off by the rain; young specimens are convex ②, but become flatter and broader ①, and are 5–10cm wide; the margin is often, but not always, striped, especially in fully developed specimens. Gills are compact and white. Stem is narrow and white, with a ring that is smooth on the outside ④; the base broadens abruptly into a volva ③.

TYPICAL FEATURES

The ring is smooth on the outside ④, making it easy to distinguish from the edible species of Amanita.

FLESH: The white flesh has no smell, and does not discolour when cut.

DISTRIBUTION: In coniferous and deciduous woodland, but mainly under deciduous trees on sandy soil.

EDIBILITY: The Panther Cap is very poisonous. Severe psychological and narcotic symptoms, including hallucinations, cramps, blurred vision and paralysis begin very soon after consumption. These symptoms are similar to those produced by the Fly Agaric, but in this case they can be fatal.

Grey Spotted Amanita ○

Amanita excelsa | *Amanita spissa*

APPEARANCE: Cap is pale grey to greyish-brown; initially convex to hemispherical, but later becomes flatter and broader ①; 5–12cm wide; skin is covered with the greyish-white to brownish fluffy remnants of the veil ②, but these may be washed off by the rain. Gills are dense and white. Stem is initially also white, but turns grey or brownish in older specimens; below the ribbed ring, the stem is slightly woolly ③, with a volva at the base ④.

TYPICAL FEATURES

Unlike the Panther Cap, the stem of the Grey Spotted Amanita gradually blends into the volva. The ring has prominent vertical ridges, and the cap has a smooth margin.

FLESH: The flesh is white and smells of radish or turnips; it discolours to brown rather than red and is grey directly beneath the cap. The flavour is mild, but eating is not recommended.

DISTRIBUTION: Mainly in coniferous woodland, but occasionally also in deciduous woods.

EDIBILITY: Edible but of poor quality. Easily confused with the Panther Cap and should therefore only be collected by experts.

67

Tawny Grisette

Amanita fulva

APPEARANCE: Cap of the young mushroom is egg-shaped or hemispherical, later broadening and becoming flattened or domed ①, 3–8cm across, thin brown to reddish-brown flesh with furrowed striations apparent on the paler rim of the cap ②. Gills are white, and free of the stem. Stem is narrow, whitish, sometimes slightly fluffy, bulbous at the base, and always surrounded by a membranous volva ③.

FLESH: Flesh is white, delicate and brittle. Smells and tastes mild.

DISTRIBUTION: In damp deciduous and coniferous woods; common.

TYPICAL FEATURES

The ringless stem and deeply furrowed rim of the cap distinguish the edible Grisette from the poisonous Death Cap (p62).

EDIBILITY: Edible when cooked, but not particularly recommended. Can cause mild poisoning if eaten raw.

SIMILAR SPECIES: The cap of the Grisette *(Amanita vaginata)* ④ is greyer and the stem is slightly banded.

Fawn Pluteus

Pluteus cervinus

APPEARANCE: Hemispherical to bell-shaped cap has thin flesh. The pale brown cap is 7–14cm in diameter when fully expanded ①; it is shiny in wet weather. Gills are free of the stem, and white in young mushrooms, later turning pink ② and then a bright fleshy red colour. Stem is narrow and fleshy, with narrow, brown, vertical stripes on a pale background ③.

FLESH: Very soft, white flesh. Smells slightly of radish and has a mild, sometimes slightly bitter, taste.

TYPICAL FEATURES

The gills on fully developed Pluteus are red and free of the stem. This is a reliable feature for identification.

DISTRIBUTION: The species is saprophytic, growing on the dead wood of deciduous and pine trees ③, sometimes in large numbers.

EDIBILITY: Fawn Pluteus is edible, but does not have a particularly good flavour and does not grow in large numbers.

SIMILAR SPECIES: Brown-flecked Winter Mushroom *(Pluteus nigroflocculosus)* ④ is also edible and the gills are dark brown in cross-section. The surface of the gills is white.

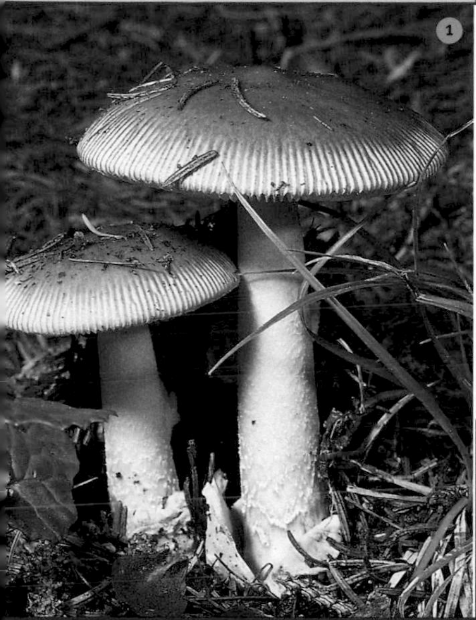
1

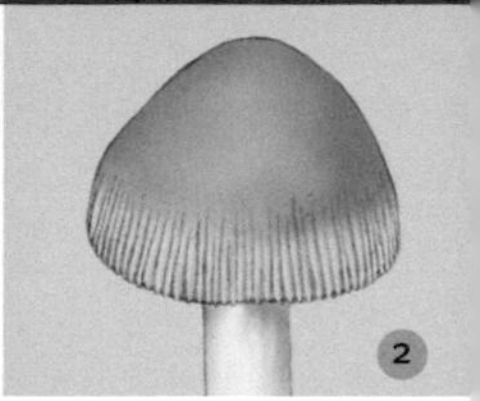
2

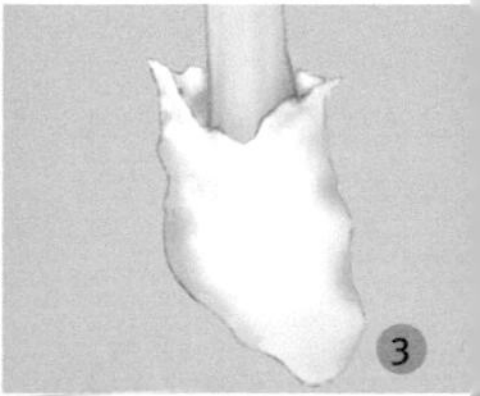
3

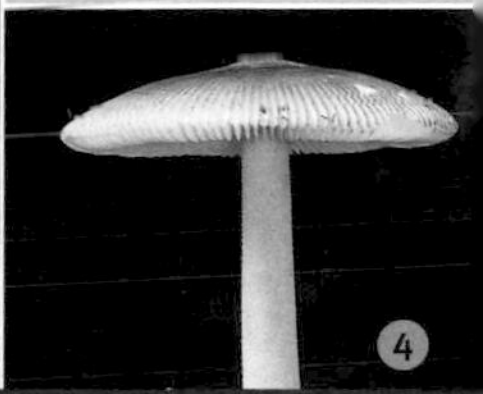
4

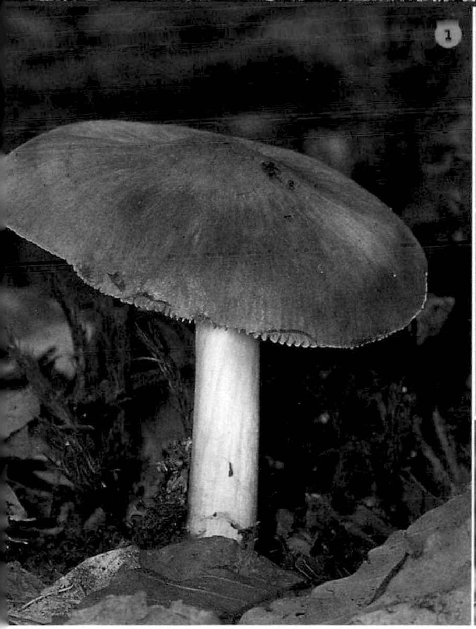
1

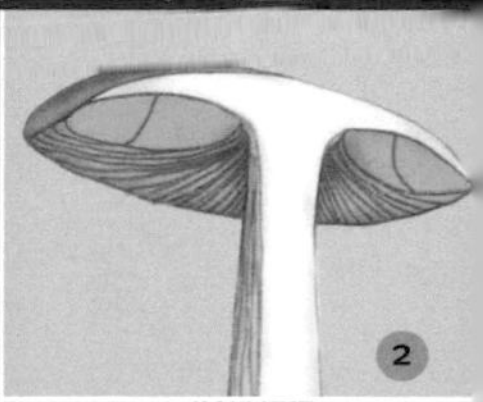
2

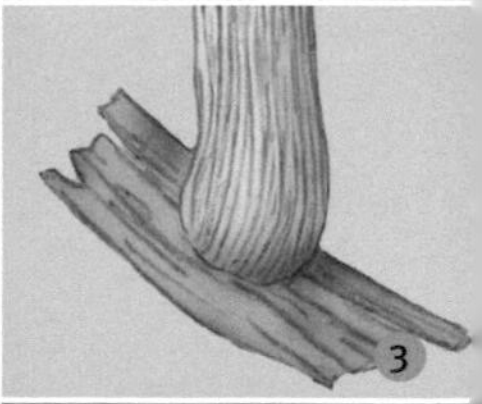
3

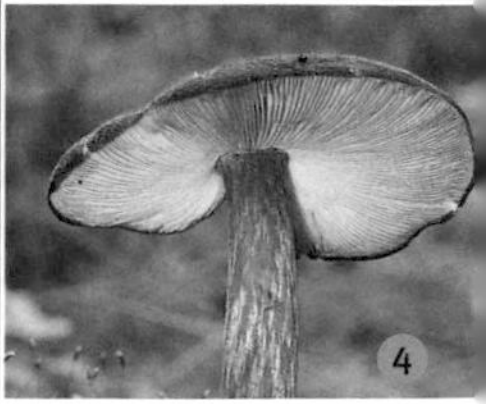
4

St George's Mushroom ¶¶¶
Calocybe gambosa

APPEARANCE: Thick flesh under the cap which is broadly convex initially, later becoming broad and flat and reaching a diameter of 10cm; creamy-white ①, but sometimes also brownish or ochre-yellow ④; surface of the cap is matte and dry, margin inrolled. Gills are compact and white ②, slightly sinuate at the stem. Stem is cylindrical, white or cream, with vertical stripes and scales.

FLESH: Flesh is also creamy-white ③ and smells and tastes floury.

DISTRIBUTION: Grows in grassland, gardens, parks and deciduous woodland; occasionally found in spruce clearings. Often forms fairy rings.

EDIBILITY: St George's Mushroom appears around St George's Day (23 April) and is a popular edible mushroom.

TYPICAL FEATURES

The uniform cream-white colour of the fruiting body and the strong smell of flour make St George's Mushroom, which appears in spring, easy to identify.

Deadly Fibrecap ☠
Inocybe erubescens

APPEARANCE: Cap has a diameter of 3–8cm and is conical and white when young, later turning straw-yellow ①, and reddening at the margin when mature ④. The surface typically has radiating fissures and is umbonate in the centre; gills are initially whitish, but turn grey-brown, and are a bright brick red, like the cap, on older mushrooms. Gills sinuate at the stem; stem is fleshy, and angular at the base ③; in young mushrooms it is white, but later turns reddish.

FLESH: Young flesh is white but turns brick red with increasing age ②. Fragrance is initially fruity, but turns unpleasantly sickly.

DISTRIBUTION: Deciduous woodland, parks and gardens; not common, but is found in large numbers in certain areas.

EDIBILITY: The large quantities of muscimol in species of *Inocybe* can cause serious or even fatal poisoning.

TYPICAL FEATURES

The flesh turns brick-red when damaged. Also, look for the radiating pattern of fissures on the surface of the cap, and the cracked cap margin.

White Fibrecap

Inocybe geophylla ☠

APPEARANCE: Cap is initially egg-shaped, later conical or bell-shaped with a prominent hump ①; usually pure white, but can be straw-yellow or purple ②; only 2–3cm broad. Surface is silky; gills are compact and slightly sinuate at the stem. Gills are white on young mushrooms, but earthy-brown when mature, whatever the colour of the cap ③. Stem is hollow ④ and often curved, white with a fluffy surface texture and longitudinal stripes; the base is not bulbous, and is often yellowed.

TYPICAL FEATURES
The gills are only white on young mushrooms. Whatever the colour of the cap, the gills of older mushrooms are always earthy-brown.

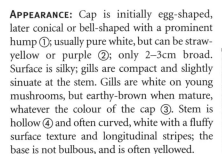

FLESH: The flesh smells of fresh bread dough.

DISTRIBUTION: In coniferous and deciduous woodland, but also commonly found beside footpaths and roads.

EDIBILITY: Like many other Inocybes, White Fibrecap contains large amount of the nerve toxin muscimol. Consumption of White Fibrecap can lead to fatal cases of poisoning.

Split Fibrecap

Inocybe rimosa ☠

APPEARANCE: Cap is a pale straw colour or grey-brown and has dense, thin stripes; young mushrooms are acutely conical ①, but deep fissures appear as the cap broadens and flattens ②; 4–6cm wide; gills are compact and sinuate at the stem; initially almost pure white, but later pale grey and finally olive-brown. Stem is cylindrical, whitish or brownish, and paler than the cap; surface is slightly shiny.

TYPICAL FEATURES
Can be recognised by the deep cracks at the cap margins, which also extend towards the centre of the cap when mature.

FLESH: Whitish, smells of dough.

DISTRIBUTION: Split Fibrecap is common in coniferous and deciduous woodland, but also grows in parks.

EDIBILITY: Like many species of Inocybe, it is very poisonous, due to its high muscimol content.

SIMILAR SPECIES: Frosty Fibrecap (*Inocybe maculata*) ④ can be recognised from the white-grey, fluffy remnants of the veil ③ in the centre of the cap. The veil remnants may be washed off by heavy rain. The species is commonly found in woodland and grassland.

73

Clouded Funnel Cap ○

Clitocybe nebularis

APPEARANCE: Cap is pale to ash grey ①, fleshy, dry, and inrolled at the margins when young; initially hemispherical, but later flat and broad, with a diameter of up to 15cm; gills are white to cream and decurrent along the stem ④; stem is full and fleshy, becoming increasingly spongy with age; grey-brown with vertical stripes ③, often slightly thicker at the base.

FLESH: The white flesh has a sour smell.

DISTRIBUTION: This mushroom is found in late autumn in coniferous and deciduous woodland; always forms fairy rings.

TYPICAL FEATURES

A reliable identification feature are the compact, brittle gills ②, which can easily be detached from the flesh of the cap with a finger.

EDIBILITY: The mushroom can be eaten if thoroughly cooked, but it can have some ill effects. Young fruiting bodies in particular can cause stomach complaints.

Wood Blewit ♈♈♈

Lepista nuda

APPEARANCE: Purple cap is up to 12–15cm in diameter ①; turns slightly brown with age, and turns paler if the soil is dry; surface is smooth and dry; gills are purple and sinuate; stem is cylindrical, and normally purple like the cap, with a fibrous, vertically furrowed surface; occasionally broadens and becomes club-shaped at the base.

TYPICAL FEATURES

The bright purple gills never turn brown so they are an easy form of identification.

FLESH: Purple ②, turning paler with age; has a sweet aromatic fragrance and taste.

DISTRIBUTION: Grows in large numbers in deciduous and coniferous woods, normally in fairy rings or rows.

EDIBILITY: This abundant mushroom is edible. Due to its distinctive flavour, it is best not to cook it with other mushrooms. Raw consumption is not recommended.

SIMILAR SPECIES: The Blewit *(Lepista saeva)* ④ is also edible. The cap is pale grey or cream, and it is normally only purple on the stem ③. The gills are yellowish.

75

Common Funnel Cap
Clitocybe gibba

APPEARANCE: Cap is 5–8cm wide; initially flat but increasingly funnel-shaped with age. Almost always has a small umbo in the centre ①. Pale brown or ochre-coloured; cap margin is inrolled, and later often has a wavy contour; gills are compact, pale orange-brown to creamy white, very decurrent at the stem ④; young stem is white, but turns ochre-yellow; spongy on the inside.

FLESH: The white flesh smells of bitter almonds and has a mild flavour.

TYPICAL FEATURES
Unlike some of its close relations, the stem of the Common Funnel Cap is always pale in colour.

DISTRIBUTION: In coniferous and deciduous woodlands; grows in clusters.

EDIBILITY: The mushroom is not classed as poisonous and can be eaten with a warning. It is not recommended as an edible mushroom, however.

SIMILAR SPECIES: Other Funnel Caps are a similar colour, but with a darker stem. Tawny Funnel Cap *(Lepista inversa)* ②, only grows in coniferous woods and has yellow-orange flesh ③. It is not edible.

Fool's Funnel
Clitocybe rivulosa

APPEARANCE: Cap is only 2–4cm wide, initially white but later a brown, fleshy colour; older mushrooms are slightly concave like a plate ①; surface is slightly frosted; gills are compact and white, turning creamy yellow in older mushrooms ③, slightly decurrent at the stem. Stem is narrow and relatively long; initially pure white but increasingly flesh-coloured with age; stem becomes hollow over time ②.

FLESH: The white flesh has a pleasant, slightly floury small.

TYPICAL FEATURES
The pale, brown cap with white frosting is typical. The surface often forms dark, circular patches as it ages.

DISTRIBUTION: Mainly grows on meadows, beside footpaths, sometimes in gardens and on grassland; less common beside forest footpaths.

EDIBILITY: The Fool's Funnel contains the alkaloid muscimol and is very poisonous.

SIMILAR SPECIES: Ivory Clitocybe *(Clitocybe dealbata)* ④ also contains muscimol in large quantities and has already caused cases of fatal poisoning. Never try to taste this mushroom.

77

he Miller

Clitopilus prunulus ⚪

APPEARANCE: Cap is 5–12cm broad with thick flesh and grey-white colour ①; initially hemispherical, then broadly convex; when mature has a central funnel; gills are creamy-white; flesh is pink or brown and very thin; very decurrent at stem ④; stem short and white, often wrinkled, and felted at the base.

FLESH: The soft, mellow flesh is whitish. It smells and tastes strongly of flour.

DISTRIBUTION: Common mushroom, found in both deciduous and coniferous woods.

TYPICAL FEATURES

The Miller's gills are very soft and friable and are not as elastic as the similar, but poisonous white Funnel Caps.

EDIBILITY: Edible, but very easy to mistake for the poisonous Funnel mushrooms (p76). Only experts should try eating The Miller.

SIMILAR SPECIES: The Frosty Funnel (*Clitocybe phyllophila*) ② is very poisonous and often found in clusters in autumn, in deciduous and mixed woodland. The gills are adnate ③.

White Domecap

Lyophyllum connatum

APPEARANCE: Grows in dense clusters; caps are white, or grey when saturated with water, and up to 10cm wide; margin is initially inrolled, but cap has irregular wavy edges when mature ①; gills are compact and white or creamy white, adnate at the stem ②, and later slightly decurrent; stem is thin and initially white, but yellowish when mature, with faint vertical furrows on the surface.

FLESH: White and very brittle; smell is sour; tastes of flour.

TYPICAL FEATURES

The creamy white gills and the characteristic rancid and sour smell are important distinguishing features of White Domecap.

DISTRIBUTION: In damp, grassy locations, beside streams and gravel paths in deciduous and coniferous woods; often grows in large numbers.

EDIBILITY: The mushroom contains substances which can cause genetic damage, so it is inedible.

SIMILAR SPECIES: Grey-gilled Funnel Cap (*Lyophyllum fumosum*) ④ is grey-brown and grows in clusters in meadows and on heathland. It is good edible mushroom. Several fruiting bodies form on a single stem (

78

easy Tough Shank

ollybia butyracea

APPEARANCE: Cap is rounded but later flattens out ①; it is about 6cm wide, with thin flesh, and is reddish, chestnut brown. The colour is paler when dry ③; surface is smooth and shiny. Gills are pure or dirty white, and sinuate; the rim has curved edges. Stem is tubular, and the same reddish-brown colour as the cap; prominent vertical furrowing; base is slightly thicker with a noticeable white felted surface ④.

FLESH: The whitish flesh can hold a lot of water and has a gristly texture. It has a pleasant smell and mild flavour.

DISTRIBUTION: Coniferous and deciduous woods; often forms fairy rings.

EDIBILITY: This fungus is edible, but does not taste particularly good.

SIMILAR SPECIES: *Collybia butyracea* var. *asema* ② is a variety of the Greasy Tough Shank; it has a grey cap and stem, but can also be found in pale grey or blackish-brown.

TYPICAL FEATURES
The chestnut-brown cap with a greasy and shiny surface is an important distinguishing feature.

Russet Tough Shank

Collybia dryophila

APPEARANCE: Cap is flat or saucer-shaped, about 3–5cm wide, thin flesh, yellowish to orange-brown ①, always slightly darker when wet. Gills of young mushrooms are whitish, turning pale creamy-yellow; compact ② and sinuate; Stem is the same colour as the cap, hollow and smooth right down to the base.

FLESH: Watery, pale cream-coloured flesh has a pleasant, slightly sour smell but little flavour.

DISTRIBUTION: The fruiting bodies of this common saprophyte are found in coniferous and deciduous woodland. Almost always grows in groups but rarely in dense clusters.

EDIBILITY: This Tough Shank is edible, but does not taste particularly good, and should therefore be cooked with other mushrooms.

SIMILAR SPECIES: The inedible, but not dangerously poisonous, Hairy Tuft *(Collybia hariolorum)* ④ has a hairy stem ③ and is noticeable for its unpleasant smell of cooked cabbage.

TYPICAL FEATURES
Russet Tough Shank has a smooth stem. The mushroom grows in groups, but not dense clusters.

Wood Woolly Foot

Collybia peronata

APPEARANCE: Cap is a pale, reddish-brown with thin flesh, broad and flat ①, 3–6cm wide. The cap is normally dry and matte, with a margin that curves slightly downwards. Gills are noticeably thick and widely-spaced, ochre-yellow, and normally sinuate ②. Stem is pale, yellow-brown, with prominent hairs at the base or on the lower half ④; slightly wider at the base, with a white or yellowy felted surface.

FLESH: The stringy, yellow-white flesh smells pleasant, but tastes acrid.

DISTRIBUTION: A saprophytic fungus that grows in coniferous and deciduous woodland, often in very large numbers.

EDIBILITY: Inedible due to its acrid taste.

SIMILAR SPECIES: Dark Rooting Shank (*Collybia obscura*) ③ has a dark purple-brown cap and stem. Almost no smell and mild flavour.

TYPICAL FEATURES

The flesh of Wood Woolly Foot initially tastes mild, but becomes acrid after a couple of seconds. The burning sensation does not last long.

Lilac Bonnet

Mycena pura

APPEARANCE: Cap broadly convex or flat; about 2–4cm wide with thin flesh that is transparent at edges, with striped appearance caused by visible gills ③; very variable colour, from pale bluish-purple to pinkish-purple ① or white to steel-grey. Gills are white to bright-pink, relatively widely-spaced and sinuate ②. Stem is thin and smooth, hollow and tubular ②; colour normally similar to that of the cap, slightly felted at the base.

FLESH: The watery white or grey-purple flesh smells strongly of radishes.

TYPICAL FEATURES

The hollow stem is felted at the base, and the strong radish smell and purple colouring are features that help identify this Mycena.

DISTRIBUTION: Common in coniferous and deciduous woods.

EDIBILITY: Lilac Bonnet was long considered to be edible, but it has been found to contain slight traces of the toxin muscimol. It is therefore now classified as mildly poisonous.

SIMILAR SPECIES: Pink Bonnet (*Mycena rosea*) ④ is larger and can be distinguished by its pink cap.

Bonnet Mycena
Mycena galericulata

APPEARANCE: Cap is 4–6cm wide and campanulate ①, though occasionally flat and broad with a wrinkled surface; the colour is a pale fleshy-brown or white. The gills are very widely-spaced, initially dirty white, but turning grey to pale pink with increasing age ②. Stem is very thin, up to 10cm long; colour is similar to the cap, very smooth and shiny, slightly felted at the base.

TYPICAL FEATURES
The pointed, campanulate cap has a distinctively wrinkled surface with numerous forks radiating from the margin to the centre.

FLESH: Greyish-white, smelling slightly mealy.

DISTRIBUTION: Mostly grows in clusters, and rarely individually, on the stumps of deciduous and coniferous trees.

EDIBILITY: Bonnet Mycena is not poisonous, but is not recommended as an edible mushroom.

SIMILAR SPECIES: Winter Bonnet (*Mycena tintinnabulum*) ④ is much smaller and has a brown cap. It appears between November and March on the stumps of deciduous trees. The stem has ragged hairs at the base ③.

Milk-drop Mycena
Milking Bonnet | *Mycena galopus*

APPEARANCE: Cap is campanulate and only 1–2cm in diameter; the grey-brown surface has prominent radiating furrows ③. Gills are white, widely spaced and sinuate. Stem is very thin ①, smooth, the same colour as the cap; the base is white and felted.

TYPICAL FEATURES
Milk-drop Mycena is named after the milky sap which seeps from the damaged stem and flesh ④. This is an unmistakable feature of the species.

FLESH: The delicate, brittle flesh is grey-white and has hardly any smell. A milky sap is exuded from damaged areas.

DISTRIBUTION: Very common; grows in coniferous and deciduous woodland. Mycena are small fungi that grow on dead wood or in the soil, normally among mosses.

EDIBILITY: Milk-drop Mycena is not poisonous, but like all Mycenas, it is not an edible mushroom.

SIMILAR SPECIES: Red Edge Bonnet (*Mycena rubromarginata*) ② is also grey-brown, but is easily distinguished by the bright red edges of its gills. The flesh releases no milky sap.

85

Fairy Ring Champignon 🍴

Marasmius oreades

APPEARANCE: Cap has thin flesh, and is 2–5cm in diameter; initially campanulate ①, but later flattening ②, though a prominent umbo remains in the centre; ochre to reddish-brown, paler in dry weather; prominent furrows at the margin. Gills are widely-spaced, creamy-yellow ③ and sinuate. Stem is 4–8cm tall, narrow, and often paler than the cap; sometimes even white ④.

TYPICAL FEATURES
The smooth, narrow stem of the Fairy Ring Champignon is tough and can be bent without breaking.

FLESH: The white flesh smells of freshly sawn wood, and has a pleasant, mild flavour.

DISTRIBUTION: Fairy Ring Champignon forms in large clusters and fairy rings early in the year, mainly in meadows, gardens and parks, at woodland edges and in fertile grassland. It is very common.

EDIBILITY: Despite its thin flesh, Fairy Ring Champignon is a highly recommended edible mushroom. Only the caps are eaten.

Garlic Fairy Cap 🍴

Marasmius scorodonius

APPEARANCE: Cap has thin flesh, pale ochre-brown ①, only 1–2cm wide, with an uneven, wrinkly, slightly striped surface; becomes broad and flat with age, often with irregular waves. The gills are wide-spaced, very broad and white ②, sinuate where they grow to join the stem ③. Stem is red-brown, 3–6cm long; normally darker than the cap, smooth and hollow ③.

TYPICAL FEATURES
The fleshy brown cap is normally darker in the centre than at the margin ①.

FLESH: The white flesh smells strongly of garlic and tastes quite acrid.

DISTRIBUTION: Exclusively in coniferous woods; the small mushrooms grow between spruces and pines on a bed of fallen needles.

EDIBILITY: The thin flesh of the cap is used as a condiment. The garlic aroma is not lost in cooking, nor when the mushroom is dried and crushed.

SIMILAR SPECIES: Oak Fairy Cap (*Marasmius querceus*) ④ grows in autumn on fallen leaves. It is inedible.

ethyst Deceiver
...ccaria amethystea

APPEARANCE: Cap is 3–5cm wide with thin bright purple flesh ①, but sometimes paler when dry; mature mushrooms have an irregularly wavy cap with a slight dip in the centre, and an inrolled margin; surface covered in fine scales. Gills are thick, bright violet in colour ②; very widely spaced with shorter sub-gills forming at the margin between the main gills. Stem is narrow and the surface has fine, vertical striations; colour is like that of the cap.

FLESH: The tough, violet-coloured flesh has almost no aroma and a mild flavour.

TYPICAL FEATURES
The shape, appearance of Amethyst Deceiver and its bright purple colour make it hard to mistake. The mushroom turns pale grey when dried.

DISTRIBUTION: Common in coniferous and deciduous woodland.

EDIBILITY: Edible, but does not grow in large numbers.

SIMILAR SPECIES: Bicoloured Deceiver *(Laccaria bicolor)* ④ is reddish-brown in colour and the base of the stem is pale blue ③.

The Deceiver
Laccaria laccata

APPEARANCE: Cap is initially broadly convex, but later becomes slightly funnel-shaped; up to 5cm wide; very variable colour, from leather brown ② to pink or a pale fleshy red ①; paler when dry; surface of cap covered in fine scales. Gills are pink or fleshy red and widely-spaced ①; usually adnate but less frequently decurrent. Stem is tough and the same colour as the cap, or slightly darker; surface has faint vertical striations.

FLESH: The flesh is reddish, occasionally paler, with a pleasant smell and a mild flavour.

TYPICAL FEATURES
The Deceiver is easy to identify by the characteristic reddish colour of the thin cap.

DISTRIBUTION: In damp coniferous and deciduous woodland; damp locations; common.

EDIBILITY: Edible, but has a low yield.

SIMILAR SPECIES: Scurfy Deceiver *(Laccaria proxima)* ④ is also edible but slightly larger. It also has bigger scales on the cap ③ and the stem has more prominent, floccose striations. Mainly grows in damp habitats and moo‐

Entoloma

loma sinuatum

APPEARANCE: Cap is fleshy, initially broadly convex, but later flat and up to 15cm wide; margin is always slightly inrolled; colour is silver-grey to ivory ③. Gills are initially creamy-yellow or yellow, and later pale-pink to red ②; they detach easily from the flesh of the cap and are slightly sinuate ④. Stem is very thick, cylindrical or club-shaped, and roughly 12cm tall; yellowish-white, with vertical striations ①.

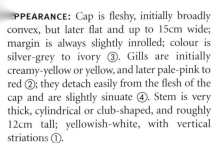

FLESH: The white, firm, mild-tasting flesh has a strong but pleasant smell of flour.
DISTRIBUTION: This fungus loves the warmth and is found in beech and mixed woodland. Often forms fairy rings.
EDIBILITY: Livid Entoloma is poisonous, and despite its pleasant taste, it can cause serious and long-lasting intestinal disorders shortly after consumption. No fatal cases of poisoning have been recorded.

Bitter Poisonpie

Hebeloma sinapizans

APPEARANCE: Cap initially broadly convex, but later flat and up to 15cm wide; margin is inrolled on young mushrooms, but curves upwards on older specimens; skin of cap is a pale yellow-brown ① to reddish-brown; slimy when wet, but has a white frosting when dry. Gills are compact, pale ochre to brown in colour, with pale edges and sinuate ②. Stem is hollow and slightly thicker at the base; white or brownish, with scaly surface ③.

TYPICAL FEATURES

Where the cap joins the hollow stem, an inverted cone forms, hanging into the hollow space ②.

FLESH: White, with a bitter taste and a strong smell of radish.
DISTRIBUTION: Can be found in large numbers, mainly under beech trees in deciduous woods; less common in coniferous; often forms fairy rings.
EDIBILITY: The mushroom causes serious digestive problems.
SIMILAR SPECIES: Poison Pie *(Hebeloma crustuliniforme)* is slightly paler but is also poisonous and also smells of radish. Often grows under lime (linden) trees, but also beside footpaths in woods. It releases tiny water droplets from the gill edges ④.

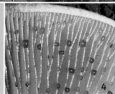

91

False Chanterelle

Hygrophoropsis aurantiaca

APPEARANCE: Cap is funnel-shaped ①, 2–7cm wide; bright orange-yellow, less frequently a russet colour; margin very inrolled, even when mature ③; skin of cap is silky with thin furrows. Gills are compact and very decurrent; initially the same colour as the cap ②, but becoming paler with age. Stem is cylindrical, slightly tapering at the base and the same colour as the cap and gills.

TYPICAL FEATURES
Unlike the Chanterelle (p138), which has a white flesh, False Chanterelle has orange flesh and true gills.

FLESH: Orange-coloured ④, relatively soft and elastic; almost no fragrance and mild flavour.

DISTRIBUTION: A common mushroom in autumn, mainly in coniferous woods. Mostly grows in the soil, but sometimes also on rotten wood.

EDIBILITY: False Chanterelle is not poisonous, but it can cause an upset stomach so it is not recommended as an edible mushroom.

Spike Cap

Slimy Spike | *Gomphidius glutinosus*

APPEARANCE: Cap is 8–12cm wide; initially hemispherical ④, but later slightly funnel-shaped; margin is always inrolled ①; grey-brown to a fleshy ochre. Decurrent gills are initially white, but turn grey with age ③, and relatively widely spaced. Stem is thick and whitish; like the cap, it is covered in a slimy layer; base of stem is bright chrome-yellow ②.

TYPICAL FEATURES
A distinguishing feature is the slimy coating on the grey-brown to ochre-brown cap, which can easily be peeled off.

FLESH: Except at the base of the stem, the flesh is white; it has almost no smell and a mild flavour.

DISTRIBUTION: Spike Cap lives in a symbiotic relationship with conifers, mainly spruces, normally in small clusters. It is quite common.

EDIBILITY: Spike Cap is edible, and some collectors value it for its flavour. The slimy skin and mucilage should be removed before cooking.

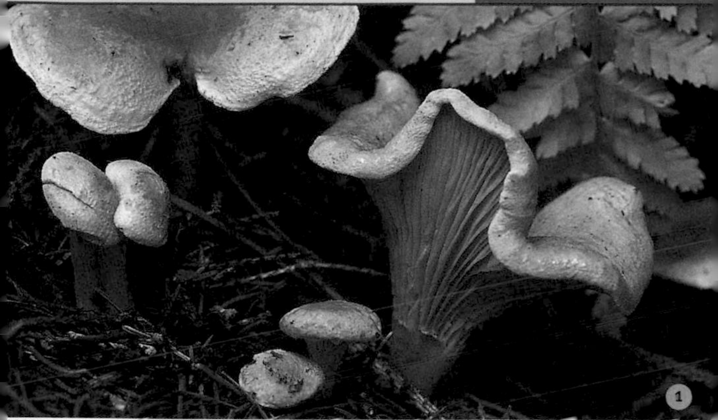
1

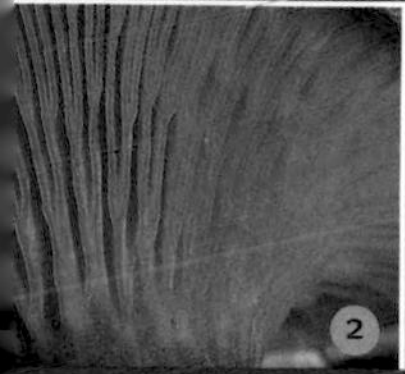
2

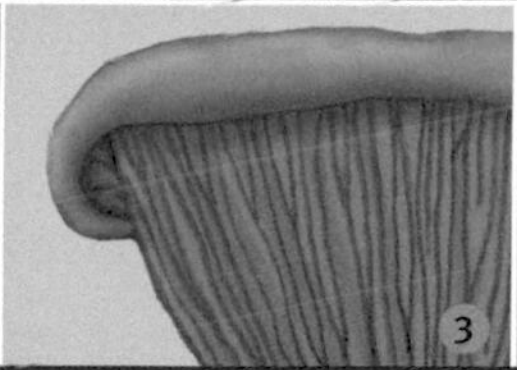
3

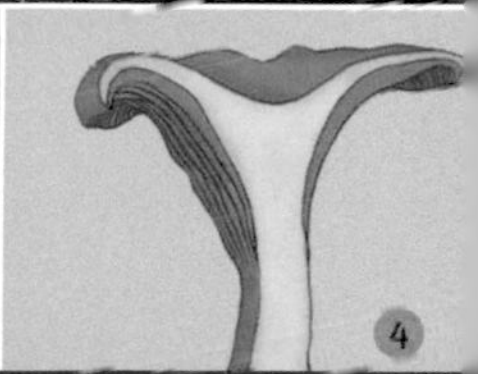
4

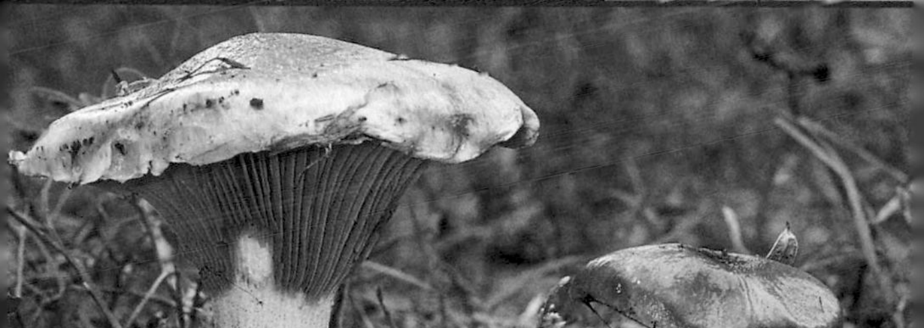
1

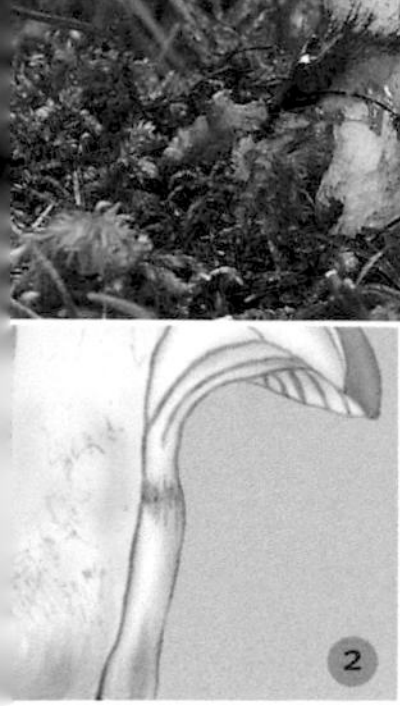
2

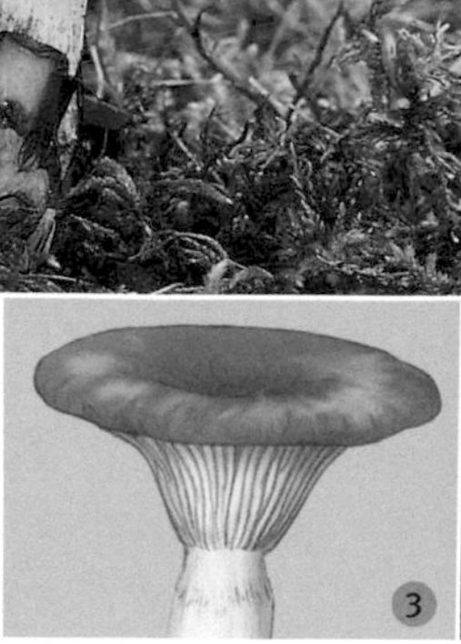
3

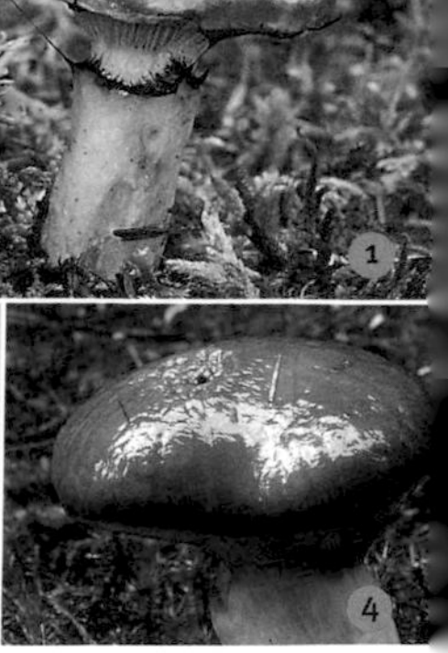
4

Velvet Roll-rim

Paxillus atrotomentosus

APPEARANCE: Cap is dark ochre to pinkish-brown ①; hemispherical when young ④, but expanding with age; up to 15cm in diameter; inrolled edge, even when mature ③; normally an irregular shape, sometime also funnel-shaped; surface is dry and silky. Gills are creamy-yellow to white, relatively compact, and decurrent. Stem is very short and thick in relation to the cap; dark brown and downy ②.
FLESH: Soft and watery; yellowish in colour; faint smell, tastes slightly sour or bitter; flesh of the stem very firm.
DISTRIBUTION: In damp woodland on dead conifer stumps. Normally grows in clusters.
EDIBILITY: This mushroom is fairly distinctive. Although it is not poisonous, it is only edible after it has been cooked for a long time, and even then it does not taste very good.

TYPICAL FEATURES
The very thick, downy, dark brown stem of Velvet Roll-rim often grows off-centre from the cap.

Brown Roll-rim

Paxillus involutus

APPEARANCE: Cap is flat and broadly convex, becoming broader and increasingly funnel-shaped ①; up to 15cm wide; ochre to rust-brown; margin is always curved and inrolled ②; in damp weather and when mature, skin of cap is greasy. Gills are a dirty yellow colour, decurrent at the stem ③. Stem is cylindrical and relatively short, about the same colour as the cap; turns brown where it is touched.
FLESH: Dirty yellow in colour; turns brownish-red when cut; slightly sour smell.
DISTRIBUTION: Very common in Europe, growing in coniferous woods, but also in other woodland types, on moors, in gardens and parks.
EDIBILITY: Highly poisonous when raw and found to cause liver damage if eaten regularly, even cooked.
SIMILAR SPECIES: Alder Roll-rim *(Paxillus filamentosus)* ④ is slightly smaller, with prominent scales. It only grows under alders.

TYPICAL FEATURES
The yellowish gills become flecked with dark reddish-brown when pressed. They detach easily from the cap.

95

Purple Wax Cap
Rosy Wood Waxcap
Hygrophorus erubescens

APPEARANCE: Cap has maroon or purplish-red spotting on a white background ①; can be completely white or even purple; broadly convex or flat and broad; often irregular and bent when mature; 7–10cm wide; thick flesh; surface is greasy. Gills very thick and widely spaced; initially white but develop maroon flecks. Stem is white; can have maroon spotting, particularly where stem joins the cap ④.

TYPICAL FEATURES
The slightly bitter flesh slowly turns yellow when it is pressed.

FLESH: White but yellows when pressed; almost no smell; slightly bitter taste.
DISTRIBUTION: In coniferous woods, mainly under spruces; not particularly common; grows in large clusters.
EDIBILITY: Slightly poisonous.
SIMILAR SPECIES: Orange Wax Cap (*Hygrophorus pudorinus*) ② is inedible and mainly grows under firs. The orange flesh is yellow at the base of the stem ③ and tastes and smells of turpentine.

Olive Wax Cap
Hygrophorus olivaceoalbus

APPEARANCE: Cap 3–6cm wide; grey-brown to olive-brown; waxy and shiny; initially rounded and inrolled, but later broad or slightly funnel-shaped. Gills are very thick and widely spaced; white to cream-coloured. Stem is white, with greyish-brown reticulation.

TYPICAL FEATURES
In wet weather, the cap and stem are very slimy ③. The surface of the stem is reminiscent of snakeskin ②.

FLESH: The soft flesh is white, without a scent and with a mild taste.
DISTRIBUTION: Grows in mossy areas in coniferous woods, particularly under spruces; mainly in mountain regions.
EDIBILITY: The mushroom is edible, but does not have a high yield. Its waxy surface makes it slightly unappetising.
SIMILAR SPECIES: The Ivory Slime Cap (*Hygrophorus eburneus*) ④ is white, grows in deciduous woodland, and is sometimes found in large numbers. Its very slimy cap feels sticky, even when dry.

97

Blackening Wax Cap
Hygrocybe nigrescens

APPEARANCE: Cap 2–4cm wide; yellow, yellow-orange or reddish-orange in colour ①, increasingly blackens with age. Young specimens are acutely conical, but later expand; cap margin is often cracked when mature ③. Surface of cap has a radiating pattern of stripes. Gills are white or yellow and free; they blacken when pressed. Stem is narrow and hollow ②, yellow to orange-yellow in colour; often with vertical black striations when mature, turning black from the base up ③.

FLESH: Flesh initially white or yellow but, like the surface of the cap, blackens with age.

TYPICAL FEATURES
The conical cap has a prominent colour and the blackening flesh is a distinguishing feature to help identify the species. Old fruiting bodies are often very black ④.

DISTRIBUTION: Widespread on grassy patches in well-lit woodland, at woodland edges and in meadows.

EDIBILITY: Can cause stomach upsets. Not much is known about its edibility, but it is classified as mildly poisonous.

Brown Mottlegill
Panaeolus foenisecii

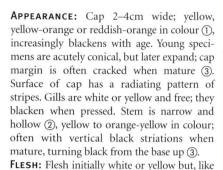

APPEARANCE: Cap up to 3cm wide; young mushrooms are campanulate ①, later becoming flatter and broader ③; when wet, a dark reddish-brown, drying to pale grey. Gills on young mushrooms are a pale olive-brown, with flaky edges; they darken with age, eventually to a rusty brown colour. Stem is long, thin and hollow ②, reddish-brown, but paler than the cap; cap has a silky sheen.

FLESH: Thin and soft, brown in colour.

TYPICAL FEATURES
After rain, Brown Mottlegill is one of the commonest mushrooms on grassy areas. Also found in urban environments.

DISTRIBUTION: Grows on freshly mown meadows, mainly in pastures and in parks; often seen after rain.

EDIBILITY: Mottlegills are not edible mushrooms. Some species contain narcotic substances which have not been sufficiently researched.

SIMILAR SPECIES: There are about 15 species of Mottlegill, some of which can only be differentiated microscopically. All grow on dung. Ringed Mottlegill *(Panaeolus fimiputris)* ④ grows on old cow pats in mountainous regions. It is the largest Mottlegill, and the only species that has a ring.

99

Conifer Tuft

YYY

Hypholoma capnoides

APPEARANCE: Caps of young mushrooms are hemispherical and a honey-yellow colour ②. When mature, they are broadly convex or expanded and up to 6cm wide. The cap is a darker orange-brown in the centre. Gills are compact, initially white, turning brownish-grey and slightly sinuate ③. Stem is narrow and hollow, paler near the top and rusty brown at the base, with fine vertical striations; grows in a tight cluster from a single base ①.

TYPICAL FEATURES

Identifying the grey-brown gills ④ of the Conifer Tuft is the only sure way to distinguish it from the inedible Sulphur Tuft.

FLESH: Flesh is yellowish-white and has almost no smell. It has a mild flavour. The flesh of the stem is brown and very tough.

DISTRIBUTION: This species grows in clusters all year round, but mainly in autumn, and only on dead conifers. Rare in the United Kingdom.

EDIBILITY: The mushroom is edible and very tasty and the only species of Tuft which is suitable for consumption. Only the caps should be eaten.

Sulphur Tuft

Hypholoma fasciculare

APPEARANCE: Cap is a bright sulphur-yellow; much darker orange-brown in the centre ①. About 5–6cm wide and completely smooth on top, but slightly floccose at the margin ③. Gills greenish-yellow to grey-green, very crowded and sinuate ②. Stems are sulphur yellow, hollow and grow from a single point in a large cluster; on young mushrooms, the stem often has a flaky ringed area beneath the cap.

TYPICAL FEATURES

Sulphur Tuft is conspicuous due to the dense clusters that grow on dead wood. Its greenish-yellow or grey-green gills are also distinctive.

FLESH: The sulphur yellow flesh has a pleasant smell, but tastes very bitter. Eating is not recommended.

DISTRIBUTION: The conspicuous fruiting bodies grow in large clusters, mainly on the stumps of deciduous trees, and less frequently on conifers.

EDIBILITY: The mushroom can cause stomach complaints, but has not been proven to be seriously poisonous.

SIMILAR SPECIES: Brick Tuft *(Hypholoma sublateritium)* ④ also grows on deciduous wood, and has a reddish colour.

Velvet Shank

Flammulina velutipes

APPEARANCE: Cap is honey-coloured to reddish-brown, and normally much paler at the margin ①. Initially broadly convex, then broad and flat, often somewhat uneven and 2–5cm wide. Skin is very smooth and slightly sticky. Gills are white to pale yellow ②, quite widely spaced, and adnate or very slightly sinuate. Stem is tough; several stems grow together from a single point; initially yellow but soon turning dark brown ③; very velvety texture.

FLESH: The flesh has a pleasant smell and a mild flavour. It is slightly yellow in colour.

TYPICAL FEATURES
This winter mushroom can be distinguished from the dangerous Funeral Bell (p54) by its white gills and velvety, ringless stem. The Funeral Bell also grows over winter.

DISTRIBUTION: A saprophyte that appears in late autumn and winter, growing mainly on deciduous wood ④; less frequently on conifers; it has been cultivated.

EDIBILITY: Velvet Shank is one of the few delicious species of edible mushroom that can be collected during the winter.

Oyster Mushroom

Pleurotus ostreatus

APPEARANCE: Cap is shell-shaped with thin flesh. It is up to 15cm in diameter and is usually a pale grey-brown to a dark steely blue, and occasionally brown. Inrolled margin, stem excentric ③. Gills are white to cream-coloured ④, very decurrent at the stem. The very short stem is whitish with felted hairs at the base ②. Several fruiting bodies almost always grow from a single point, growing densely on top of each other ①.

FLESH: The white flesh becomes tough with age and has a fibrous structure. Smell and taste are mild and pleasant.

TYPICAL FEATURES
The colour and shape of the Oyster Mushroom may vary, but it is hard to confuse with other mushrooms due to its distinctive shape and growing habits.

DISTRIBUTION: The Oyster Mushroom grows in late autumn or winter and is a saprophytic or parasitic fungus, mainly on beeches, poplars and other deciduous trees; occasionally found on spruces.

EDIBILITY: This delicious species has long been cultivated as an edible mushroom and is easily available from supermarkets.

103

Broad-gilled Agaric

Megacollybia platyphylla

APPEARANCE: Cap is initially hemispherical but later flattens ① to a diameter of 10–15cm; margin of older mushrooms has an angular, cracked edge; colour is dark brown, grey-brown or pale grey, often with striations. Gills are white and thick and very sinuate. Stem is white with vertical striations.

FLESH: White; normally devoid of smell or taste but sometimes slightly bitter.

DISTRIBUTION: This saprophytic species grows on rotting tree stumps or small pieces of wood on the ground. Common in both coniferous and deciduous woodland.

TYPICAL FEATURES

Even if the fruiting bodies appear to grow from the soil, the tough, root-like hyphae ④ are always attached to rotten wood.

EDIBILITY: Unpleasant taste so not recommended for eating.

SIMILAR SPECIES: Plums and Custard *(Tricholomopsis rutilans)* ② has a reddish, flaky, scaly cap ③ and only grows on dead conifer wood. It also has an unpleasant taste, so is not edible, despite the name.

Rooting Shank

Xerula radicata

APPEARANCE: Cap is conical at first, but later expands ① up to 8cm wide, often with a broad central umbo ③. Very thin flesh; greyish-brown, reddish-brown or yellowish-brown; very slimy ④ and wrinkled; when damp, the margin is translucent and furrowed. Gills are pure white ④, widely spaced, and sinuate. Stem is very tough and conspicuously long and narrow; white near the top and brown lower down.

TYPICAL FEATURES

The skin of the cap is very greasy. The thin stem has a root-like extension ②, which can often extend into the soil to a depth of more than 10cm.

FLESH: The white flesh has almost no smell and a mild flavour.

DISTRIBUTION: Found individually or in small groups in deciduous and mixed woodland, mainly beneath beeches; also on beech roots and stumps.

EDIBILITY: Rooting Shank is not poisonous, but quite tough and it does not have a pleasant flavour.

SIMILAR SPECIES: Porcelain Mushroom *(Oudemansiella mucida*, pictured p14) is white or grey, and has a ringed stem. It grows on dead beeches.

Common Ink Cap
Coprinus atramentarius

APPEARANCE: Cap is initially ovate ②, later campanulate ①, up to 7cm wide and a whitish grey to pale grey-brown in colour, with prominent, radiating ridges. Margin curls upwards when mature ④. Gills are compact and initially white-grey, but later almost black. Stem is white and up to 9cm long; cylindrical, hollow and slightly thicker at the base ③.

FLESH: The white, mild-tasting flesh has almost no smell.

DISTRIBUTION: Deciduous woodland, beside footpaths; meadows, gardens and parks. Always grows on heavily manured soil.

TYPICAL FEATURES
The mushroom has a short life-span and characteristic features are the radiating ridges on the cap and the inky secretion from the mature cap and gills.

EDIBILITY: Young mushrooms are delicious. The mushroom, however, contains coprine, which blocks the breakdown of alcohol in the body. Under no circumstances should alcohol be consumed, even in very small amounts, for two days after consumption of the mushroom.

Weeping Widow
Lacrymaria lacrymabunda

APPEARANCE: Cap is 4–10cm wide, but sometimes up to 20cm wide, with thin flesh; initially convex, but later flat and broad ①. Fibres hang from the margin; these are the remnants of the veil that connects the cap and stem of the young mushrooms ③; skin of the cap is ochre or yellow-brown, with a flaky and scaly texture. The gills are brown ④, with darker mottling, turning almost black; sinuate where they join the stem. Stem is

TYPICAL FEATURES
Young specimens of Weeping Widow mushroom are especially prone to releasing tiny water droplets.

cylindrical and brittle, hollow, with a distinctive flaky and scaly surface; colour similar to that of the cap; paler near the top; below the cap there is often a darker ring of fibres ②.

FLESH: Brown flesh has almost no smell and a slightly spicy flavour.

DISTRIBUTION: This mushroom is common in places, growing in groups beside roads and footpaths, in fields, meadows, gardens and parks, on nitrogen-rich soil.

EDIBILITY: Weeping Widow is edible but not recommended.

Bitter Web Cap

Cortinarius infractus

APPEARANCE: Cap is olive-green to yellow-brown or pale brown; 6–12cm in diameter and irregularly domed ①, sometimes also umbonate; skin is slimy when damp, but shiny when dry; margin is inrolled and normally straight, but on mature mushrooms it may be uneven. Gills olive-brown ②, compact, and sinuate at the stem. Stem is cylindrical or slightly club-shaped, often with a bulbous base; whitish-grey to olive or ochre brown.

TYPICAL FEATURES

A distinguishing feature of the species is the sooty, olive-brown gills, which become darker with age ②.

FLESH: The white or blue-grey flesh often tastes bitter, but may also have a mild flavour. It has no distinct smell.

DISTRIBUTION: Deciduous and coniferous woods; mainly in mountains.

EDIBILITY: Bitter Web Cap is inedible.

SIMILAR SPECIES: Olive Web Cap (*Cortinarius subtortus*) ④ is also inedible and is a paler olive-yellow colour. It has bright olive-coloured gills ③ and a narrow club-shaped stem.

Girdled Web Cap

Cortinarius trivialis

APPEARANCE: Cap is initially rounded but later becomes broadly convex and 4–10cm wide, with a central umbo; olive-yellow ① to pale honey-brown ③ in colour; slimy when wet, but shiny when dry. The sinuate gills are compact and blue-brown, turning rust-brown with age. Stem is up to 10cm tall, narrow and slimy; the upper stem is covered in remnants of the floccose veil connecting the cap margin to the stem of the young mushroom ②.

TYPICAL FEATURES

The stem is covered in a layer of grey slime, which peels off when dry and then resembles snake skin ①.

FLESH: Firm; pale ochre in colour; mild smell and taste.

DISTRIBUTION: Girdled Web Cap is very widespread and is found in deciduous woods and, less frequently, in coniferous woods with limestone soil.

EDIBILITY: Not edible.

SIMILAR SPECIES: Blue-stemmed Web Cap (*Cortinarius collinitus*) ④ is mainly found in coniferous mountain woodland and is covered in a bluish-purple, sticky mucous layer. The stem is not banded.

Deadly Web Cap

Cortinarius orellanoides

APPEARANCE: Cap is initially conical, but becomes broadly convex or broad and campanulate with age, always with a central umbo ③; cap is 3–8cm in diameter, colour is yellowish, orange or red-brown. The sinuate gills are thick and widely spaced; they are the same colour as the cap ④. Stem is long and normally cylindrical ①, but sometimes slightly club-shaped and the same colour as the cap. The surface is banded with yellow fibrous flakes ②.

FLESH: The red-brown flesh smells slightly of earth or radish.

DISTRIBUTION: In coniferous woods and boggy or marshy ground. Not found in the United Kingdom.

EDIBILITY: Contains the poison orellanine, which damages the kidneys. Consumption of the mushroom has caused serious poisoning.

TYPICAL FEATURES

The orange-brown cap with fine scales has a pointed and sometimes more rounded umbo, even when mature ③.

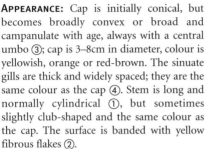

Orange Web Cap

Cortinarius orellanus

APPEARANCE: Cap is flat or broadly convex when mature, 3–8cm wide, often with a wide umbo; orange or fox-brown with a flaky, felted skin. Gills are broad, thick and widely spaced, with a cinnamon-brown or orange colour ①, sinuate where they join the stem. Stem is long and sturdy, tapering slightly at the base; paler than the cap ③, with a matte, slightly fibrous surface.

FLESH: The flesh smells of radish and is a bright orange-yellow colour ②.

DISTRIBUTION: Prefers warmer regions and is mainly found in deciduous woodland and under pines. It is relatively rare.

EDIBILITY: The poison in this mushroom can cause serious, often untreatable kidney damage, which can lead to death in a couple of months. Symptoms do not begin until one or two weeks after consumption.

TYPICAL FEATURES

The orange-yellow flesh ② has a very strong radish aroma. Do not taste the mushroom under any circumstances.

111

Goliath Web Cap ○
Cortinarius praestans

APPEARANCE: Cap of the young mushroom is purple-brown ④, slimy, and covered in a white, membranous veil; when mature can measure up to 20cm in diameter; reddish-brown colour with a wrinkled margin ①. Gills are initially grey, later turning brown and adnate or slightly sinuate. Stem is white, thick and bulbous ②; full and fleshy; often covered with the remnants of the veil.

FLESH: Flesh is a pale lilac colour with a firm texture; has almost no smell and a mild taste.

DISTRIBUTION: The large fruiting bodies can be found in warm deciduous woodland and on lime soils. Normally grows in clusters, and sometimes in fairy rings.

EDIBILITY: Good edible mushroom, but rare in the United Kingdom.

TYPICAL FEATURES
Young fruiting bodies are covered in a silky, blue-white veil. Ochre-yellow remnants of this veil later cover the broad cap of the mushroom ③.

Contrary Web Cap 🍴
Cortinarius varius

APPEARANCE: Cap is pale yellow- or fox-brown and is initially rounded, but becomes flat and broad when mature ①; 4–10cm wide; skin of cap is often very slimy. Gills are compact and sinuate where they join the stem; they are a blue-purple colour ②, but later turn a pale ochre or cinnamon-yellow colour. Stem is white, with a flaky ringed zone; the base of the stem is broad and bulbous, but has no edges.

FLESH: The white flesh has no smell and a mild flavour.

DISTRIBUTION: This Web Cap is common and lives in symbiosis with trees. Mainly found in coniferous woods, under spruces.

TYPICAL FEATURES
The stem is white and bulbous at the base with a floccose ring near the top. This turns brown when mature, due to the spores having been released ①.

EDIBILITY: Edible with a pleasant flavour.

SIMILAR SPECIES: Yellow Web Cap *(Cortinarius delibutus)* ④ is also inedible and is yellow on the stem beneath the flaky ringed zone ③.

113

Gassy Web Cap

Cortinarius traganus

APPEARANCE: Cap of young specimens is convex ①, later expanding, and 8–12cm in diameter; colour is a bright purplish-violet, becoming a paler, dirty ochre-brown with age ④; skin of the cap is silky and shiny, and not greasy. Unlike the cap, the gills are never lilac, but always ochre or saffron-brown ③, and sinuate where they join the stem. Stem is thick and bulbous or club-shaped; initially a pale violet, becoming brown with age at the base; it

TYPICAL FEATURES
The saffron-yellow flesh of the Gassy Web Cap normally has a very pungent smell.

has a floccose ring zone, which turns brown when the spores are released.

FLESH: Flesh has a pungent smell and is ochre or saffron-yellow in colour.

DISTRIBUTION: Grows in clusters in coniferous woods, less often in deciduous woodland.

EDIBILITY: This fungus is mildly poisonous and causes stomach upsets.

SIMILAR SPECIES: Goatscheese Web Cap *(Cortinarius camphoratus)* ② is similar; young specimens have purple gills; has a sickly smell.

Violet Web Cap

Cortinarius violaceus

APPEARANCE: Cap is initially hemispherical, later becoming broadly convex or flat, and up to 15cm wide; fleshy and dark blue-violet in colour ①, turning blackish-brown with age; skin of the cap has fine scales ③ or is silky and dry. Gills are thick and widely spaced; sooty bluish-violet in colour ④. Stem is dark purple, but can turn slightly brown in colour around the ring zone when the spores are released; stem surface is slightly scaly and the base is often bulbous and thick ②.

TYPICAL FEATURES
All parts of the mushroom are a bright bluish-violet and the flesh smells strongly of cedar wood.

FLESH: Blue-violet and smells of cedar wood; unpleasant taste.

DISTRIBUTION: Violet Web Cap is quite widespread but is rare. Found in coniferous and deciduous woodland and often forms fairy rings.

EDIBILITY: Normally classed as edible, but tastes of cedar wood and is not recommended.

115

Yellow Knight

Tricholoma equestre

APPEARANCE: Cap up to 10cm wide; greenish-yellow ① and slightly paler and yellower at the margin; skin of cap is shiny, and slightly greasy when damp. Gills are thin and crowded; sulphur yellow. Stem smooth and yellow; often slightly rounded or bulbous at base.

FLESH: Flesh is white and only yellow just beneath skin ④. Smells strongly of flour and has a mild taste.

DISTRIBUTION: Coniferous woods, mostly under pines; less common in deciduous woods.

TYPICAL FEATURES

Yellow Knight is rare. Its distinguishing features include its strong floury smell, and bright yellow, crowded gills.

EDIBILITY: Consumption is not recommended. Eaten regularly, substances in the mushroom can cause fatal inflammation of the heart muscles.

SIMILAR SPECIES: Sulphur Knight *(Tricholoma sulphureum)* ②, is slightly poisonous and has an unpleasant smell of coal gas. It has widely spaced gills and the flesh is sulphur yellow throughout ③.

Soap-scented Tricholoma ☠

Tricholoma saponaceum

APPEARANCE: Cap 5–10cm wide; varies in colour from dirty white, to yellow-green, to brownish or reddish, to grey-black; in darker specimens the margin is always paler ①; skin of cap is smooth or scaly, and normally matte. Gills are white to greenish-yellow ②, sinuate at the stem. Stem is cylindrical, sometimes slightly bulbous, white to cream-coloured, with a smooth or slightly scaly surface ③. There is a darker variety of this mushroom ④, which has a stem with almost black scales.

TYPICAL FEATURES

The colour of the cap can vary greatly, and the most distinguishing feature of the mushroom is the flesh, which smells of soap, and slowly turns red.

FLESH: A strong, soapy smell is characteristic of the white flesh, which can redden slightly. The taste is usually mild but sometimes slightly bitter.

DISTRIBUTION: Coniferous and deciduous woodland. Fruiting bodies normally grow in large groups.

EDIBILITY: Mildly poisonous and can cause stomach upsets. The level of toxicity was underestimated until recently.

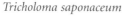

Striped Tricholoma

Tricholoma tigrinum ☠

APPEARANCE: Cap initially hemispherical, later flat and broad or broadly convex; up to 12cm wide; skin is silver-grey and flakes into woolly, pale grey or greyish-black scales ④; margin very inrolled. Gills of young specimens are white, later creamy-yellow, and sinuate. Stem is whitish, with a flaky, or slightly scaly surface.

TYPICAL FEATURES
The upper part of the thick stem often releases small water droplets ①.

FLESH: The whitish flesh smells of meal and has a mild taste.

DISTRIBUTION: Scattered in conifer and deciduous woodland; but not found in the United Kingdom. Often grows in clumps or in fairy rings.

EDIBILITY: Striped Tricholoma tastes good, but causes serious digestive problems, which can last several days.

SIMILAR SPECIES: Grey Knight (*Tricholoma terreum*) ② is very similar but edible. It does not have the striking floury smell. Its gills are ash-grey and more compact, and the cap is smooth and felted ③.

Scaly Knight

Tricholoma vaccinum ✖

APPEARANCE: Cap conical or campanulate ①, later becoming broader and normally bluntly umbonate; only 3–8cm wide; pale ochre to reddish-brown; surface very scaly; cap margin has woolly or felted veil remnants. Gills are initially whitish, later with brown spotting, and eventually completely brown; very sinuate at the stem. Stem is narrow, hollow and reddish-brown; slightly paler under the cap.

TYPICAL FEATURES
Woolly and flaky remnants of the veil hang from the margin of the very scaly cap of the Scaly Knight ②.

FLESH: The white flesh has a slightly reddish tint, which deepens when cut. It has a slightly earthy smell and a bitter taste.

DISTRIBUTION: Normally grows in clusters in coniferous and deciduous woodland, on grassy plains and beside footpaths; common in mountain regions. Rare in the United Kingdom.

EDIBILITY: Bitter and inedible.

SIMILAR SPECIES: The skin of the Matt Knight (*Tricholoma imbricatum*) ④ peels away at the cap margin ③, and the white flesh, which has a bitter, slightly grassy taste, becomes visible beneath.

119

Saffron Milk Cap   ⑂∣∣

Lactarius deliciosus

APPEARANCE: Cap 3–10cm wide, concave in the centre, with pale red to orange-red concentric circles; some parts have a green colouring; heavily frosted when dry, greasy when damp. Gills are orange-yellow, turning green when pressed. Stem is the same colour as the cap, but with darker and larger spots ①; stem is hollow in older specimens.

TYPICAL FEATURES

This Milk Cap has a characteristic carrot-orange milky sap, which turns greenish when dry. It has darker spots on the stem.

FLESH: The pale, orange-red flesh turns green after a couple of hours when broken ④. It contains a carrot-coloured milky sap and smells slightly fruity.

DISTRIBUTION: In coniferous woods under pines.

EDIBILITY: An excellent edible mushroom when cooked thoroughly.

SIMILAR SPECIES: Salmon Milk Cap (*Lactarius salmonicolor*) ② is also edible; it only grows under fir trees. It has an orange-red flesh ③. The milky sap turns dark red after a couple of hours.

False Saffron Milk Cap   ⑂∣∣

Lactarius deterrimus

APPEARANCE: Cap is initially inrolled, but later broader and always funnel-shaped ①; pale orange, often with narrow, darker concentric circles; develops large green flecks with age ④. Gills are compact and the same colour as the cap, turning green when pressed ③. Stem is cylindrical and sometimes hollow and the same colour as the cap.

TYPICAL FEATURES

False Saffron Milk Cap can be identified by its location beneath spruce trees, its fruity aroma, and its orange milky sap, which turns wine-red after a couple of minutes.

FLESH: The yellowish flesh smells fruity and tastes slightly bitter. When damaged, it releases an orange-red, milky sap, which quickly turns maroon ②.

DISTRIBUTION: In late summer and autumn, this fungus can be found in large numbers in woodland, where it lives in a symbiotic relationship with spruce.

EDIBILITY: Like all other Milk Caps with red milk, the False Saffron Milk Cap is edible, but is not nearly as good as the Saffron Milk Cap.

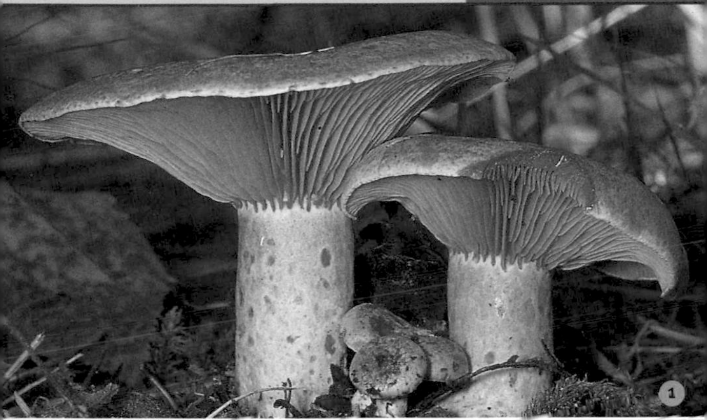
1

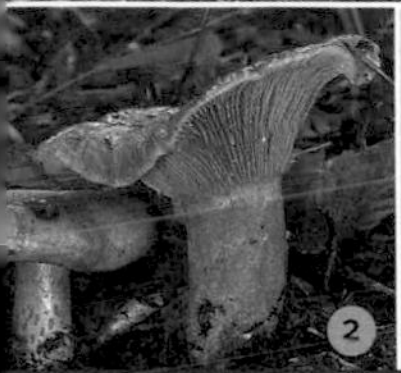
2

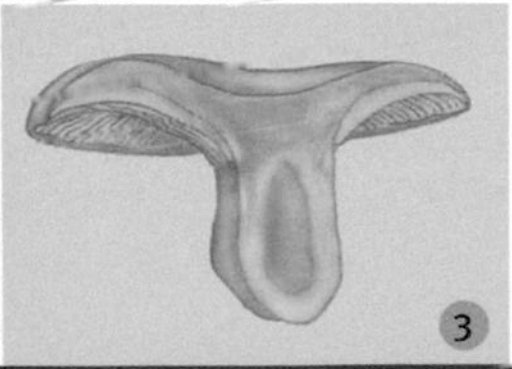
3

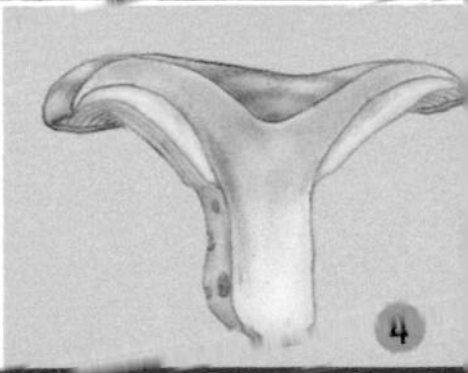
4

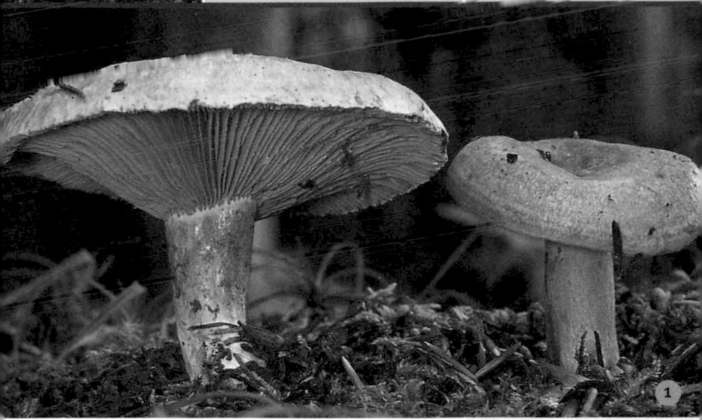
1

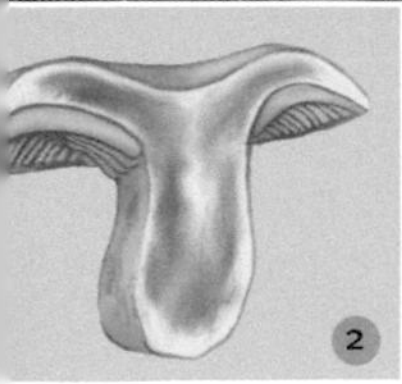
2

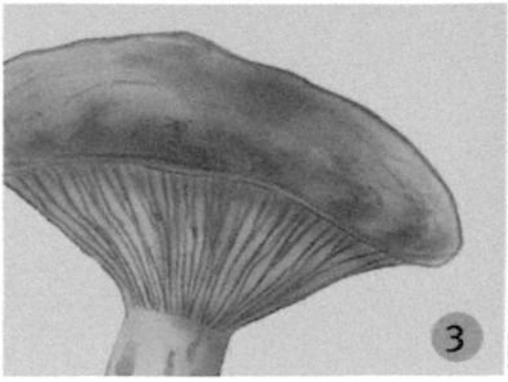
3

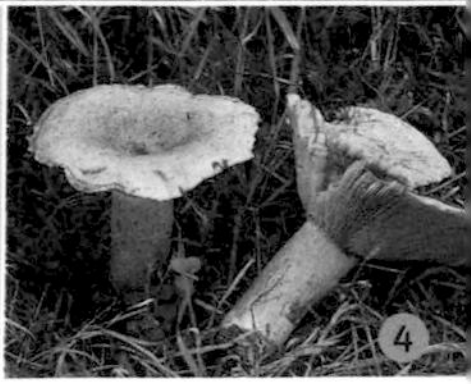
4

Dusky Milk Cap 🍴
Birch Milk Cap | *Lactarius ligniotus*

APPEARANCE: Cap is dark brown or black-brown; 2–6cm wide; margin inrolled in younger specimens, but later broader and funnel-shaped, with a pointed umbo in the centre ③; skin of the cap is silky, matt and dry. Gills are white in contrast to the cap, becoming increasingly cream-coloured with age. Stem is long and a similar colour to the cap; always wrinkled at the top ①.

TYPICAL FEATURES
Characteristic features are the very contrasting white gills and the pointed umbo in the centre of the cap.

FLESH: The white flesh turns pale pink where cut ②. Smells and tastes earthy, with a slightly bitter aftertaste. The white milk also tastes slightly bitter.

DISTRIBUTION: Mainly in coniferous forests on high ground.

EDIBILITY: A useful edible mushroom, but retains its earthy flavour.

SIMILAR SPECIES: Pitch Black Milk Cap *(Lactarius picinus)* ④ has an acrid taste and has no umbo on the cap; only found in mountain woods.

Pale Milk Cap
Lactarius pallidus

APPEARANCE: Cap of younger specimens is flattened with a slight indentation in the centre; funnel-shaped when mature; 10–15cm wide; pale fleshy red ①; shiny surface, greasy in wet weather. Gills are flesh or ochre-coloured; relatively compact. Stem is short and thick, slightly paler than the cap, turning brown when pressed; hollow when mature ②.

TYPICAL FEATURES
Characteristic features of the Pale Milk Cap are its symbiosis with beeches, and the pale colour and faint fruit smell of the gills.

FLESH: The orange-yellow flesh releases a lot of white milk, which does not discolour. Smell and taste are mild and slightly fruity.

DISTRIBUTION: This common fungus grows in deciduous woodland under beeches.

EDIBILITY: This species is edible when thoroughly cooked, but does not have a pleasant flavour and is not recommended as an edible mushroom.

SIMILAR SPECIES: Violet Milk Cap *(Lactarius uvidus)* ④ also releases large amounts of milky sap, which quickly turns purple in contact with air ③. The milk has a bitter flavour.

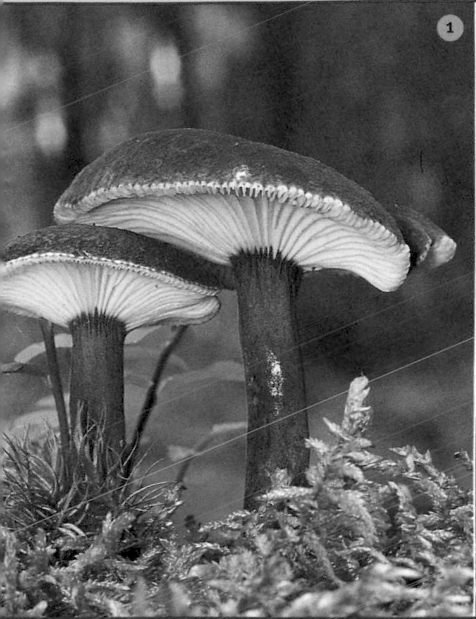
1

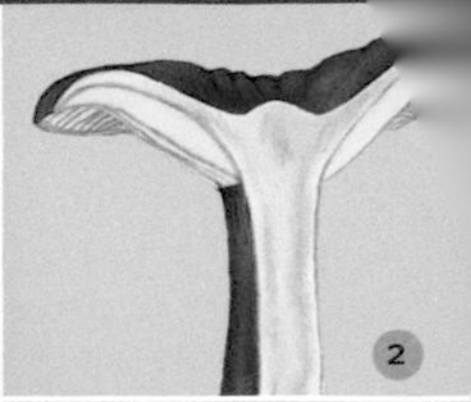
2

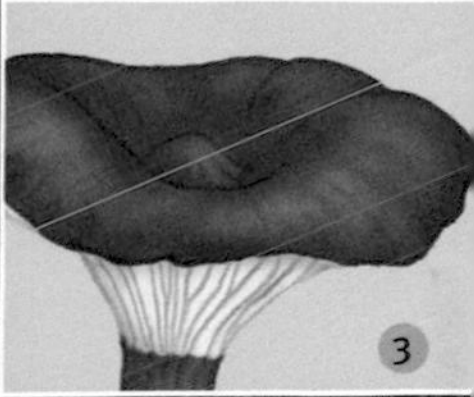
3

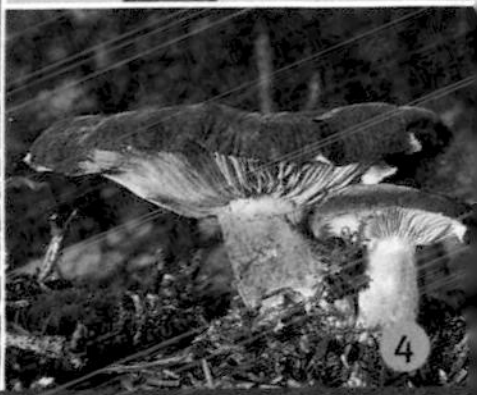
4

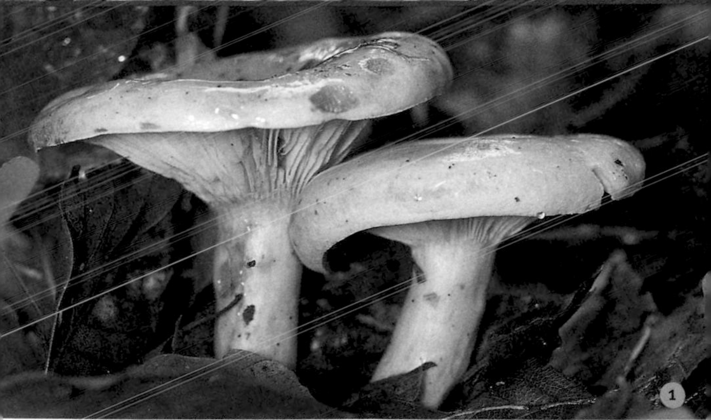
1

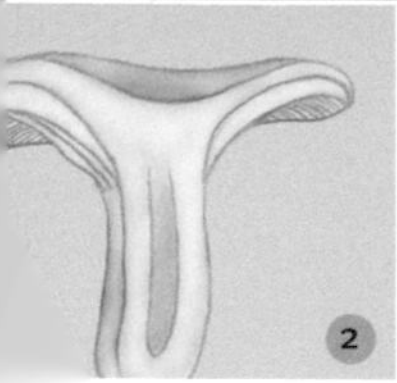
2

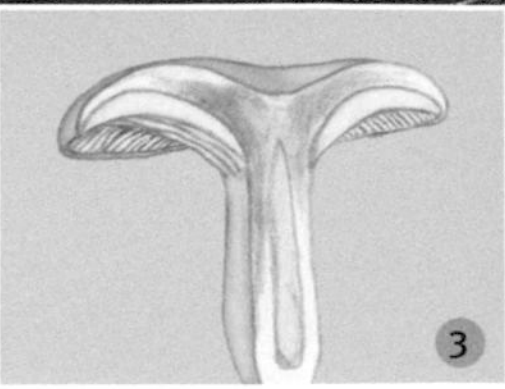
3

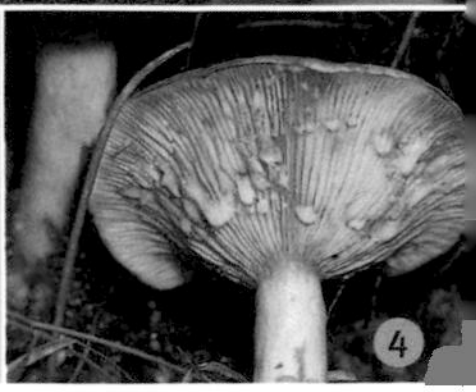
4

Rufous Milkcap

Lactarius rufus

APPEARANCE: Cap is conical when young, but broader and flatter or slightly funnel-shaped when mature ①; 5–8cm wide, normally with a pointed umbo in the centre ②; skin is reddish-brown with a silvery frosting. Gills are ochre coloured ③ to pink, adnate or slightly decurrent. Stem is cylindrical, becoming hollow with maturity; always paler than the cap.

TYPICAL FEATURES

There is almost always a small, pointed umbo in the middle of the cap ②. The red-brown skin is not shiny and has a silvery frosting.

FLESH: The flesh is white and smells strongly of resin. It eventually turns pale red ④. It releases a lot of white, bitter-tasting milky sap.

DISTRIBUTION: In coniferous woods but also under birches; common.

EDIBILITY: Normally classified as inedible due to the acrid taste of the milk. Can cause stomach upsets. In some parts of Europe, it is valued as an edible mushroom.

Tawny Milk Cap

Lactarius volemus

APPEARANCE: Cap is broadly convex or flat, with a slight funnel in the centre ②; 5–15cm wide; orange or russet, with a silky, dry surface, often with white frosting. Gills are yellowish ① and brittle, turning brown where damaged; adnate or slightly decurrent. Stem is full, thick and fleshy ③ and the same colour as the cap or paler.

TYPICAL FEATURES

Unlike the Rufous Milk Cap, the orange or russet cap of the Tawny Milk Cap has no central umbo ②.

FLESH: White to brownish; smells fishy when raw; contains a lot of white sap that turns brown as it dries.

DISTRIBUTION: Found in coniferous and deciduous woodland, where it lives beneath pines and spruces in a symbiotic relationship.

EDIBILITY: Tawny Milk Cap is normally eaten fried and it is a popular variety in many regions. Unfortunately it is becoming rare.

SIMILAR SPECIES: The common Orange Milk Cap *(Lactarius mitissimus)* ④ is reddish-brown in colour and inedible. It grows mainly in coniferou woods. Flesh is yellow to orange-brown.

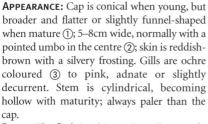

125

Funnel Milk Cap

Lactarius scrobiculatus

APPEARANCE: Cap of young specimens is hemispherical, but broader and funnel-shaped on older fungi ① and up to 20cm in diameter; margin remains inrolled ②. Skin is straw-coloured or lemon yellow, often very greasy; skins flakes off in scales at the margin. Gills are compact and creamy-white and slightly decurrent at the stem ④. Stem is short and cylindrical; yellow in colour with darker, tear-shaped pits ③.

TYPICAL FEATURES
A characteristic feature of this sharp-tasting mushroom is the numerous tear-shaped pits on the stem ③.

FLESH: The flesh has a sharp flavour but a pleasant fruity aroma and is yellowish-white in colour. The white milky sap turns sulphur-yellow in contact with the air.

DISTRIBUTION: This common Milk Cap forms a symbiotic relationship with conifers; mainly grows beneath spruces.

EDIBILITY: Inedible when raw. Can cause serious stomach upsets, even when well cooked.

Woolly Milk Cap

Lactarius torminosus

APPEARANCE: Cap on younger specimens has an inrolled margin ①, later flat and funnel-shaped in the centre; up to 10cm wide; salmon or pink-coloured with circular patches; dense, felted coating, with woolly flakes at margin ②. Gills are white to flesh pink, very compact and slightly decurrent. Stem is cylindrical and white or brownish pink in colour.

TYPICAL FEATURES
The cap margin is inrolled and is noticeably ragged and flaky ②.

FLESH: The white, firm flesh has a pleasantly fruity smell, but tastes acrid. Damaged areas release lots of white sap which is also very sharp in taste.

DISTRIBUTION: Found exclusively under birches.

EDIBILITY: The consumption of Woolly Milk Cap can cause serious stomach upsets. Nevertheless, it is a popular edible mushroom in some parts of Eastern Europe.

SIMILAR SPECIES: Bearded Milk Cap *(Lactarius pubescens)* ④ is rarer and grows under birches. The margin is less ragged and does not have coloured patches ③. Taste is acrid.

127

Fleecy Milk Cap

Lactarius vellereus

APPEARANCE: Cap is funnel-shaped ①, and very inrolled at the margin; white, developing brown spots with age; up to 20cm wide, sometimes reaching 30cm in diameter; skin is silky and woolly at the margin ③. Gills initially white but later cream; widely spaced and decurrent at the stem. Stem is white and relatively short in comparison to the cap ④.

FLESH: Flesh is white and quite firm; releases small quantities of a mild-tasting milk. Flesh itself has a sharp taste.

TYPICAL FEATURES

Young fruiting bodies of Fleecy Milk Cap half develop beneath the soil so the cap is covered in soil debris.

DISTRIBUTION: In coniferous and deciduous woodland; common everywhere, often in clusters.

EDIBILITY: Fleecy Milk Cap can cause serious stomach upsets.

SIMILAR SPECIES: Peppery Milk Cap (*Lactarius piperatus*) ② has milk that turns grey-green and a smooth cap.

Blackening Brittlegill

Russula nigricans

APPEARANCE: Cap of fully developed mushroom is flat with an inrolled margin ① or funnel-shaped; initially whitish, later earthy brown or grey-brown, and finally black; up to 15cm wide; surface is matte. Gills are thick and brittle, and widely spaced ③; dirty white to pale ochre in colour. Stem is short and thick; whitish in young specimens, and increasingly blackening in older mushrooms.

FLESH: The flesh is firm but brittle and turns orange or red when sliced ②, later blackening. The smell is faint and the taste unpleasant.

TYPICAL FEATURES

The thick, widely spaced gills do not change colour with age. They only turn black when cut ③.

DISTRIBUTION: Common in coniferous and deciduous woodland.

EDIBILITY: Inedible due to its unpleasant taste.

SIMILAR SPECIES: Crowded Brittlegill (*Russula densifolia*) ④ is also inedible and has a slightly paler grey-brown cap. The gills are more crowded.

Gilded Brittlegill ♟♟♟
Russula aurea

APPEARANCE: Cap of young mushroom is hemispherical, becoming flatter and broader ①, and up to 10cm wide; older specimens have a central funnel-shaped depression ④; surface is bright gold or lemon yellow; mature specimens have red patches ③, and are sometimes uniformly red in colour. Gills are compact and bright yellow, even in young specimens, particularly when cut. Stem is cylindrical to club-shaped; basic colour is white, with a more or less pronounced chrome-yellow frosting.

TYPICAL FEATURES
Gilded Brittlegill has noticeable yellow gills and a white stem with a yellow frosting ②.

FLESH: The firm, whitish flesh has almost no smell and a mild flavour. Flesh is yellow directly beneath the skin of the cap ④. Unlike the closely related Milk Caps *(Lactarius)*, Brittlegills do not produce any sap.

DISTRIBUTION: In coniferous and deciduous woodland, mainly at higher altitudes.

EDIBILITY: Gilded Brittlegill has a very good flavour.

Crab Brittlegill ♟♟♟
Russula xerampelina

APPEARANCE: Cap has diameter of 5–12cm; distinctive wine red or purplish-red colour ①; surface is dry and matte, only greasy when it rains. Gills are brittle and crowded and pale cream in colour ④, developing brown spots a while after they are pressed. Stem has purplish-red frosting, and also turns brown when pressed ②.

TYPICAL FEATURES
The flesh initially has no aroma, but older mushrooms and dried specimens smell strongly of fish.

FLESH: The white flesh turns brown in contact with the air ③. Ripe mushrooms have a very fishy smell, but a mild taste.

DISTRIBUTION: Crab Brittlegill is found in coniferous woods, particularly in mountain regions.

EDIBILITY: The mushroom is edible, and the fishy smell disappears during cooking. Young fruiting bodies are best for eating.

SIMILAR SPECIES: There are other Brittlegills *(Russula)* that are very similar in appearance, but taste acrid and are therefore inedible.

131

Charcoal Burner
Russula cyanoxantha

APPEARANCE: Cap is initially convex, but expands and develops a central, funnel-shaped depression; up to 12cm in diameter; variable in colour, from pale or dark violet ① to greenish ④, and occasionally also with ochre spotting; skin is shiny and greasy. Gills are pure white, thin and soft. Stem is thick and up to 10cm tall; normally white but occasionally pale violet.

TYPICAL FEATURES
While most other species of *Russula* have brittle gills, those of the Charcoal Burner are waxy.

FLESH: The white flesh has almost no smell and has a mild flavour.

DISTRIBUTION: Common. The fruiting bodies can be found in deciduous and coniferous woods, often under beeches.

EDIBILITY: One of the most delicious species of Brittlegill.

SIMILAR SPECIES: Primrose Russula *(Russula sardonia)* ② has a fruity smell but a sharp taste and mainly grows under pines. It has a similar cap colour to the Charcoal Burner, but lemon-yellow gills ③ and a purple stem.

Sickener
Russula emetica

APPEARANCE: Cap is spherical on younger specimens but becomes broadly convex or flat and broad, with a striped margin ③; up to 10cm in diameter; skin is a distinctive cherry or blood-red colour ①. Gills are white and sinuate. Stem is cylindrical and white, initially hollow, but stems on older specimens have a cottony filling ②.

TYPICAL FEATURES
The conspicuous red cap colour, in combination with the white stem and the fruity-smelling flesh make the Sickener easy to identify. The stem is never red.

FLESH: Under the skin of the cap, the white flesh is slightly reddish. It is brittle and smells of fruits, but tastes very acrid.

DISTRIBUTION: Mainly in coniferous woods, but occasionally in deciduous woods; quite common.

EDIBILITY: Inedible and causes stomach upsets as the name implies.

SIMILAR SPECIES: Sickener is very variable and can take on a variety of appearances. There is a paler and more delicate variety, known as *Russula emetica* var. *betularum* ④.

133

Brown-capped Brittlegill 🍴
Russula integra

APPEARANCE: Cap is initially hemispherical, but later saucer-shaped, with a distinct depression in the centre; skin is shiny and greasy, very occasionally dry; colour varies from purple to maroon, reddish-brown or chocolate brown; the margin of older fruiting bodies is striped ①. The brittle gills are initially whitish, turning pale or dark yellow ②. Stem is cylindrical with a cottony filling ③.

TYPICAL FEATURES
When ripe, the ochre-yellow gills contrast with the white stem.

FLESH: White and tough, but brittle. Smells fruity and has a mild flavour.
DISTRIBUTION: This species of Brittlegill is most common in mountainous areas and grows in coniferous and mixed woodland, mainly under spruces and firs.
EDIBILITY: Brown-capped Brittlegill is a good edible mushroom.
SIMILAR SPECIES: Burning Brittlegill (*Russula badia*) ④ is acrid and is easily identified by its taste. It smells strongly of cedar wood.

Ochre Brittlegill
Russula ochroleuca

APPEARANCE: Cap is lemon yellow to ochre ①, and up to 10cm wide; initially broadly convex, but later flatter and broader and slightly funnel-shaped; skin is shiny, and greasy in damp weather. The sinuate gills are compact and white, often with brown flecks when ripe. Stem is thin and cylindrical, slightly broader at the base; white but increasingly grey with age.

TYPICAL FEATURES
The stem is hollow ②. The skin of the cap can easily be peeled away at the margins.

FLESH: White; variable in flavour and may be mild, acrid or slightly bitter. The smell is faintly fruity.
DISTRIBUTION: This is the commonest of the Brittlegills. Found in coniferous woods, and lass frequently in deciduous woodland.
EDIBILITY: Ochre Brittlegill is not poisonous, but often has a sharp flavour and is not recommended for eating.
SIMILAR SPECIES: Geranium Brittlegill (*Russula fellea*) ④ often grows under beeches. The flavour is more acrid than that of the Ochre Brittlegill. Its pale, ochre-coloured cap has prominent furrows at the margin ③.

Russet Brittlegill ¶¶¶
Russula mustelina

APPEARANCE: Cap is up to 15cm wide; yellow-brown to dark brown ① and can be matte, sticky or greasy, depending on the weather. Gills are pale cream-coloured; older specimens have brown flecks; relatively crowded and sinuate. Stem is stocky; initially white but later develops a brownish colour; inside, the stem is divided into compartments ④.
FLESH: Firm, white flesh turns brown when cut. Very faint smell, sometimes slightly fishy. Mild flavour, reminiscent of hazelnuts.
DISTRIBUTION: Russet Brittlegill is only found on high ground, where it mainly grows in coniferous woods, under spruces.
EDIBILITY: A very pleasant tasting edible mushroom.
SIMILAR SPECIES: Stinking Brittlegill *(Russula foetens)* ② has a slimy, ochraceous-brown cap, with prominent marginal furrows ③. Horrible smell of putrefaction. Tastes slightly bitter, turning sickly.

TYPICAL FEATURES
The cap of the Russet Brittlegill develops beneath scattered pine needles; the skin is covered in particles of earth and needles.

The Flirt ¶¶¶
Russula vesca

APPEARANCE: Cap is initially hemispherical ②, later becoming flatter and broader, with a slight depression in the centre; up to 10cm in diameter; flesh various shades of purple ①; skin of the cap recedes slightly at the margin; greasy and sticky when damp. Gills are pure white ④, later with rust-coloured spots; sinuate at the stem, and sometimes slightly decurrent. Stem is cylindrical and relatively short; white, often with rust-brown flecking; the stem base is slightly pointed where it enters the ground ②.
FLESH: The flesh has a mild flavour, slightly reminiscent of hazelnuts; almost no smell. It is white and quite firm. It breaks quite easily.
DISTRIBUTION: Very common in coniferous and deciduous woodland, mainly on sandy soils.
EDIBILITY: A pleasant-tasting edible mushroom, one of the best of the edible Brittlegills.

TYPICAL FEATURES
The skin of the cap does not quite extend to the margin, and the white gills can be seen from above, forming a white border ③.

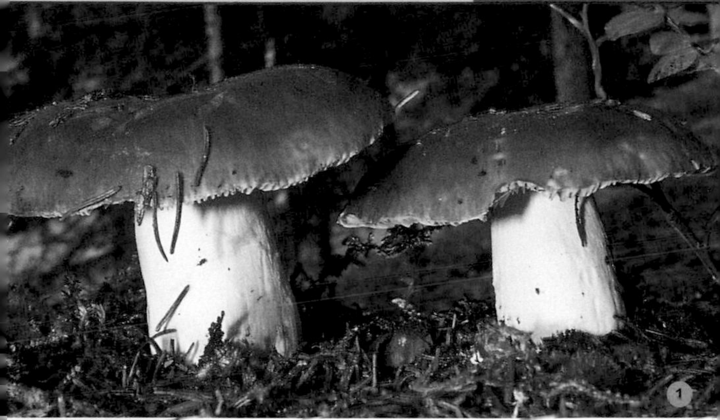
1

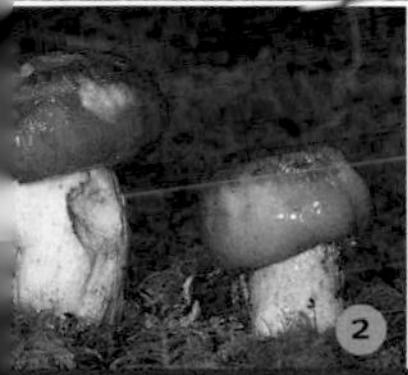
2

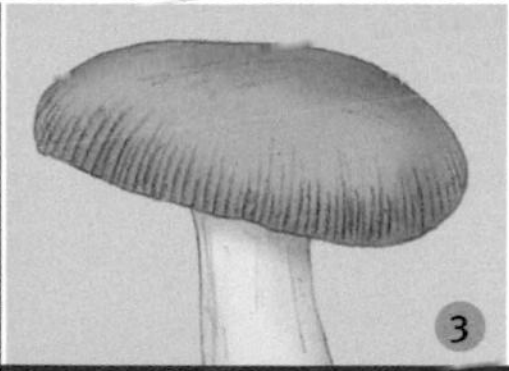
3

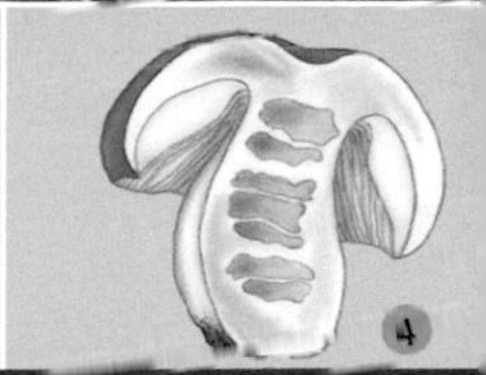
4

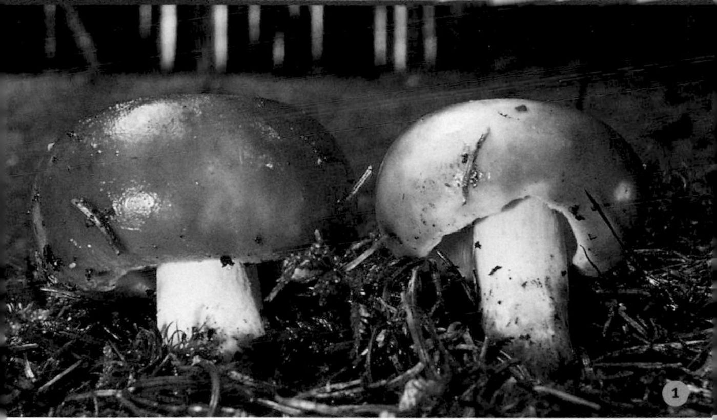
1

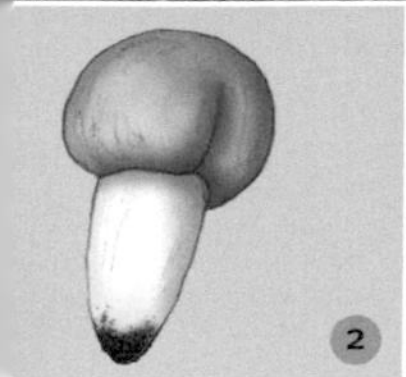
2

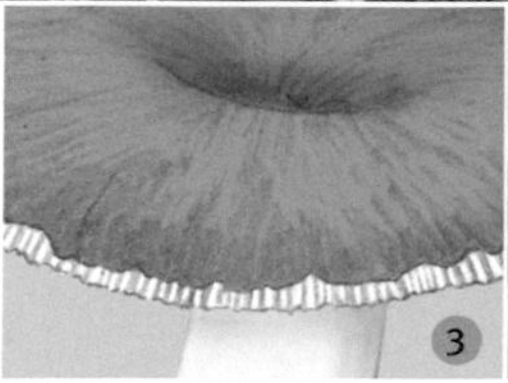
3

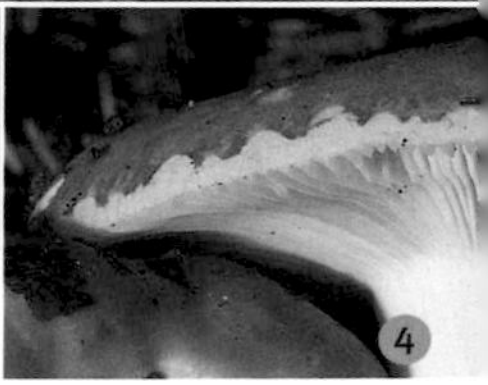
4

Chanterelle

Cantharellus cibarius

APPEARANCE: Cap is 11cm in diameter; pale to bright yellow-orange in colour; fleshy and domed, inrolled at margin; later funnel-shaped with a wavy margin ①, and becoming a paler yellowish-white; often purple in mountain regions ③. Skin is silky and dry. Beneath the cap there are 2mm-thick, forked ridges or veins the same colour as the cap, which run the whole way up the stem. Stem is fleshy ②.

FLESH: Firm flesh is white or pale yellow. Smells of apricots, but tastes mildly peppery.

TYPICAL FEATURES

The stem is funnel-shaped, narrowing at the base. The apricot colour is unmistakable as are the veins that replace gills.

DISTRIBUTION: Chanterelles normally grow under pines and spruces, less often in deciduous woodland, under birches, beeches and oaks.

EDIBILITY: Excellent and edible.

SIMILAR SPECIES: The smaller and less common Velvet Chanterelle (*Cantharellus friesii*) ④ grows in deciduous woodland and is a brighter orange-yellow in colour. It has a silky to woolly cap.

Trumpet Chanterelle

Cantharellus tubaeformis

APPEARANCE: Cap is up to 5cm wide, with thin flesh, which forms a hollow tube right down to the base of the stem ②; margin is wavy; surface is yellow-brown, grey-brown, olive or dark brown; skin is smooth or slightly wrinkled; the underside is light grey or occasionally yellow and has thick, rigid ridges of the same colour; these are joined by horizontal veins ①. Stem is narrow, hollow and a variety of yellow colours.

FLESH: The tough, white flesh smells slightly

TYPICAL FEATURES

Fully developed Trumpet Chanterelles have a hole in the centre of the cap, which extends to the base of the stem.

spicy, and occasionally slightly musty. The taste is mild.

DISTRIBUTION: Common in areas of central and northern Europe; grows in the moss in deciduous and coniferous woodland, in damp areas.

EDIBILITY: This edible mushroom is delicious but has very thin flesh. It can be pickled or preserved in vinegar.

SIMILAR SPECIES: The larger Golden Chanterelle(*Cantharellus aurora*) ④ is also edible and grows in coniferous mountain woodland. Its cap has an orange-yellow underside and is wavy or curly with tattered edges ③.

Pig's Ear

Gomphus clavatus

APPEARANCE: Young specimens have a distorted shape, later becoming club-shaped, with an incision on one side; up to 10cm high and of a uniform width; surface hairless and smooth ②, initially purple, later flesh-coloured and finally ochre-grey to ochre-yellow. Exterior is veinous and wrinkled, pale pink or purple in colour ①, sometimes also with a dusting of ochre spores. Ridges forked with horizontal banding running almost to the base.

TYPICAL FEATURES
Often grows in clusters, and the club-shaped fruiting body has gill-like purple veins on the outside.

FLESH: White, soft and delicate; no smell but tastes similar to the Chanterelle.

DISTRIBUTION: This species grows in woodland in hilly regions, under spruces, firs and beeches; rare in lowlands.

EDIBILITY: A delicious edible mushroom with a high yield.

SIMILAR SPECIES: Flat-topped Club Fungus *(Clavariadelphus truncatus)* ④ is also edible and grows in mountain woods. It has a rounded cap ③, which is reminiscent of old, club-shaped pig's ears.

Horn of Plenty

Craterellus cornucopioides

APPEARANCE: Trumpet or funnel-shaped; margin is initially inrolled, but is later lobed and cracked ②. Surface is an ashen grey ① or brownish-grey, and is almost smooth or slightly rugose. The outside of young specimens is smooth, matte and black and develops vein-like folds with age; can become whitish or pale grey when the spores are released.

TYPICAL FEATURES
The very thin flesh is always grey in colour and can easily be identified by its hollow, trumpet-shaped form.

FLESH: The grey, very thin and elastic flesh has a strong scent and a mild, slightly earthy flavour.

DISTRIBUTION: The Horn of Plenty is found in deciduous and mixed woods under beech, oak and silver fir; grows in large numbers.

EDIBILITY: The mushroom is tough when cooked, and thus only suitable for mixed dishes. Dried and ground, it makes a very good condiment.

SIMILAR SPECIES: Ashen Chanterelle *(Cantharellus cinereus)* ④ is smaller and rarer, but has a similar colour. It has ridges on the underside ③.

Wood Hedgehog
Hydnum repandum 🍴

APPEARANCE: Cap is initially convex, but later becomes flat with irregular bumps ①. Margin is inrolled and lobed and up to 15cm wide. Colour is whitish-yellow to yellow-orange; skin is dry and silky or leathery. The underside is covered with 2–6mm long, brittle, pointed spikes ②, which are slightly decurrent and the same colour or paler than the cap. Stem is short and sturdy ③, with an irregular shape; it is always paler than the cap.

TYPICAL FEATURES
The underside of the cap has compact and brittle spikes. The stem is normally positioned off-centre.

FLESH: Whitish, mellow and without scent. Tastes of oatmeal.

DISTRIBUTION: Normally on limestone soils in deciduous and coniferous woodland. Common and often found in groups.

EDIBILITY: Only young fruiting bodies taste good. Older mushrooms become bitter, but can be used for flavouring.

SIMILAR SPECIES: Terracotta Hedgehog *(Hydnum rufescens)* ④ is an orange-brown colour and smaller, with a narrower stem.

Scaly Tooth
Sarcodon imbricatus ◯

APPEARANCE: Cap is convex in younger specimens and has an inrolled margin; later flat and broad with a central depression and a wavy margin; up to 15cm wide; skin is dry, grey or brown, and has large, circular scales ④, which become smaller and more dense towards the margin. Underside has dense, brittle spikes which are initially white, but turn pale grey with age ①. These spikes are decurrent at the stem, which is excentric.

TYPICAL FEATURES
The scales on the cap of the Scaly Tooth become darker with age, and make the fungus look a dark brown colour.

FLESH: The mellow, firm flesh is dirty white to greyish-brown in colour ③; it has a spicy aroma and older specimens have a sharp or bitter taste.

DISTRIBUTION: Grows in large clusters under pines and spruces, mainly in mountainous woodland.

EDIBILITY: Young fungi can be used for flavouring when dried.

SIMILAR SPECIES: The inedible Bitter Tooth *(Sarcodon scabrosus)* has a blackish stem base ② and smaller scales.

143

Split Gill

Schizophyllum commune

Appearance: Fruiting body is mussel- or fan-shaped and 1–5cm wide. It is attached to the ground by a long shaft ③. Surface is whitish-grey with radiating forks, and it is covered in felted, woolly hairs. Turns brownish over time, and the margin is lobed or curved, wavy ② and slightly bent ④. Gills are fan-shaped ①, with a split along the edge; beige-brown to reddish or purplish-grey.

TYPICAL FEATURES

The Split Gill has greyish-white fruiting bodies with felted hairs, and is the only species in the world whose gills are split throughout their entire length.

Flesh: When damp, the ochre-coloured, thin flesh swells and becomes leathery and elastic. When it is dry, it is hard and brittle. Tastes slightly sour.

Distribution: Split Gill appears all year round and mainly grows on dead and freshly cut deciduous and coniferous wood. Occasionally also grows on damaged parts of living trees.

Edibility: This fungus is not suitable for eating.

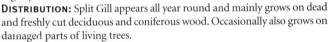

Hairy Curtain Crust

Stereum hirsutum

Appearance: Fruiting body is mussel-shaped and flexible; 1–5cm wide, and arranged in imbricated rows like roof tiles ④. The surface ② has greyish-yellow to yellow-brown coloured patches, with characteristic bristly or shaggy hairs; margin is yellow and wavy. The spore-bearing layer on the underside of the mushroom ③ is smooth or slightly bumpy, and is always hairless; pale or orange-yellow to brownish in colour.

TYPICAL FEATURES

The bristly or shaggy hairs and the yellow surface of the cap ① make Hairy Curtain Crust quite easy to identify.

Flesh: The thin flesh has a leathery and tough texture.

Distribution: This resistant bracket fungus is the first to grow on fallen wood. It is commonly found year round on the trunks of deciduous trees but less commonly on coniferous wood. It is a pest because it causes white rot in stored wood.

Edibility: Hairy Curtain Crust is inedible because of its tough texture.

Tinder Bracket
Hoof Fungus | *Fomes fomentarius*

APPEARANCE: Fruiting body is hoof or bracket-shaped ①; 10–30cm wide; initially ochre or red-brown, turning grey-brown when mature. Surface often has banding in several colours ②, with a hard, hairless crust. Underside has fine pores, which are whitish on younger specimens, but brownish on older fruiting bodies; they form in layers, which makes it possible to tell the age of this perennial fungus ③.

FLESH: Yellow-brown at the core of the cap; light and corky; fluffy and cottony; tough.

DISTRIBUTION: Always on deciduous wood, on diseased or dead beech and birch trunks.

EDIBILITY: The flesh of the core of the cap is slightly acrid and this sensation remains for a long time. Not suitable for eating.

SIMILAR SPECIES: The Willow Bracket *(Phellinus igniarius)* has a rounded and pale cap rim ④. Mainly grows on the Silver Willow.

TYPICAL FEATURES

The soft tissue in the core of the cap of the Tinder Bracket is in contrast to the hard, woody shape of the fungus.

Red-rimmed Polypore
Fomitopsis pinicola

APPEARANCE: Fruiting body has a broad, bracket-shaped base which grows directly from the substrate; young specimens are a uniform yellow-ochre, later turning greyish-black and often growing without a distinct shape or hoof-shaped; 20–30cm wide; surface is bumpy ④, with a sticky, resinous, often cracked crust. There is usually an additional white area of growth around the wavy margin, which later turns reddish or orange-yellow ③.

TYPICAL FEATURES

Red-rimmed Polypore's white underside often releases watery droplets ①.

The underside has very narrow pores, which are initially cream-coloured, turning pale ochre or lemon yellow; pores form in annual layers ②.

FLESH: Pale yellow or wood-coloured flesh is up to 3cm thick, initially corky and tough, and later hard. It has a sour smell and tastes bitter.

DISTRIBUTION: Common in mountainous regions. Grows year round on spruces, and occasionally on silver firs and alders; rarer on other coniferous and deciduous trees.

EDIBILITY: The tough texture and bitter taste of the flesh make it inedible.

Artist's Bracket
Ganoderma lipsiense

APPEARANCE: Fruiting body has no stem for several years; kidney or fan-shaped and up to 50cm wide, only 5cm thick. Surface is greyish-brown to cinnamon, and concentrically crenate; bald and uneven, with a hard, matte crust; outer growing edge is always white ①. Pores are very narrow; brown with white openings, which turn dark brown when pressed; older specimens grow in layers ③.

TYPICAL FEATURES
The grey-brown, concentrically crenate surface is often completely covered in cocoa-coloured spores ④.

FLESH: Initially whitish, becoming reddish-brown with age; wooden and corky in texture.
DISTRIBUTION: On stumps or living trunks of deciduous trees, mainly beeches; less common on coniferous trees; widespread and common.
EDIBILITY: The woody flesh of this species is inedible.
SIMILAR SPECIES: Oak Mazegill *(Daedalea quercina)* ② is also inedible and grows on the trunks of old oaks. The underside does not have rounded pores, but a labyrinthine fruit layer.

Birch Polypore
Piptoporus betulinus

APPEARANCE: Fruiting body is fan-, bracket- or mussel-shaped and about 20cm wide and 4cm thick; white specimens are white to cream-coloured, becoming ochre to greyish-brown with age ①. Point of attachment is pillow-like and often located to one side. Surface is smooth and unbanded ③. Underside is whitish, with narrow pores ④. Pores are rounded or slightly angular; white in colour.

TYPICAL FEATURES
Birch Polypore is one of the few tree fungi that will only grow on birches.

FLESH: The white, corky and tough flesh is juicy. It tastes sour and bitter.
DISTRIBUTION: Grows on ill or aging trees, eventually killing them.
EDIBILITY: The tough, rubber-like flesh makes the fungus unsuitable for eating. Used to be used as an alternative for leather in straps, and as stoppers for snuff boxes.
SIMILAR SPECIES: Cinnamon Bracket *(Hapalopilus rutilans)* ② is also inedible and commonly grows on dead deciduous and coniferous wood. It has a thick, soft, cinnamon-coloured flesh.

148

Chicken of the Woods ○
Sulphur Polypore | *Laetiporus sulphureus*

APPEARANCE: Fruiting body is fan or club-shaped and up to 40cm wide. Segments grow over each other like roof tiles ①. Bright yellow with orange-red patches and a sulphur yellow, bulging margin ④ with a slightly wavy fringe ③. Becomes paler when dry and a dirty cream colour with age ②. Upper surface is silky; underside has a more intensive colour, with narrow, sulphur-yellow pores, which release a yellow liquid in younger specimens.

TYPICAL FEATURES
The bright yellow colour and the particularly sulphur-yellow margin make the fungus easy to identify.

FLESH: The young, bright yellow flesh is very delicate and juicy. It has a pleasant smell and tastes tart and sour. Flesh of older specimens is hard, brittle and dry, and loses a lot of its colour.

DISTRIBUTION: Not fussy and grows on several species of deciduous tree, and less frequently on coniferous wood. Grows on larches in the Alps.

EDIBILITY: Young, delicate fruiting bodies can be coated in breadcrumbs and baked. They must be thoroughly washed before cooking.

Giant Polypore 🕸
Meripilus giganteus

APPEARANCE: Fruiting body consists of segments which are 8–20cm wide, and leaf-like or fan-shaped. They sit like roof tiles or rosettes on the substratum ①; can be up to 80cm wide and weigh 50kg. Surface is grainy or scaly, and yellow-brown in young specimens, later turning dark brown, but always with a paler margin; the pores run along the trunk and are round, narrow, and white in colour.

TYPICAL FEATURES
The underside of the fruiting body has black pores when pressed, or in older specimens ②.

FLESH: The fibrous mushroom has thin, tough flesh. It is initially white, becoming slightly reddish in colour, and eventually turning black and leathery. The flesh has a sour, bitter flavour.

DISTRIBUTION: Giant Polypore is a parasite, growing at the base of deciduous stumps, usually beeches, but occasionally also Silver Firs.

EDIBILITY: Not recommended as an edible mushroom.

SIMILAR SPECIES: The caps of Hen of the Woods (*Grifola frondosa*) ③ are more greyish-brown in colour, and the pores do not turn black. It can be 50cm wide ④ and grows near or on old Oaks.

151

Hairy Bracket
Trametes hirsuta

APPEARANCE: Mostly kidney-shaped or hemi-spherical, very occasionally rosette-shaped; 3–10cm wide. Surface has short hairs and forked banding ①; colour varies from white to yellow-brown to greyish-black; young specimens have a brown margin, in older specimens this is normally covered in green algae. Pores are round and dense ④ with thick walls; often decurrent at the tree; whitish-grey, becoming increasingly greyish-brown with maturity.

TYPICAL FEATURES
The hairs on the surface of this fungus have a coarse, felt-like or bristly texture.

FLESH: Inside is white and leathery and smells slightly of aniseed.

DISTRIBUTION: Common over the winter period and grows on stumps or dead trunks of various deciduous trees.

EDIBILITY: The tough flesh is not edible.

SIMILAR SPECIES: Birch Mazegill (*Lenzites betulina*) ② grows on birches and other deciduous woods, often together with the Hairy Bracket. It has a gill-like spore-bearing layer ③.

Turkeytail
Trametes versicolor

APPEARANCE: Always grows in a cluster of individual caps, mostly in rows or arranged like roof tiles ①; caps are kidney- or fan-shaped, with a wavy and sharp margin ②; 2–10cm wide. Surface is silky and shiny, with smooth and silky, multi-coloured banding, which can be black, blue, reddish, yellowish or white, or also green from algae growth. Pores are narrow and initially white ③, but later turn cream-coloured or light brown.

TYPICAL FEATURES
The very thin, tough caps have colourful, silky and shiny bands, which make the Turkeytail easy to identify.

FLESH: Thin and quite tough; pure white; has no smell or taste.

DISTRIBUTION: Turkeytail can be found year round on various deciduous trees, on dead branches and trunks.

EDIBILITY: Not suitable for eating.

SIMILAR SPECIES: The Ochre Bracket (*Trametes multicolor*) ④ appears in late autumn and spring. It is buckled where it joins the tree and has an intense brown colour. It does not have silky, shiny patches.

Conifer Mazegill

Gloeophyllum sepiarium

APPEARANCE: Fruiting body about 2–8cm wide and 1.5cm thick; semicircular, or bracket-shaped, or can grow in a rosette; broadly convex or flat where it is attached to the substrate. Surface covered in bristly hairs; multi-coloured, forked banding, initially rust-brown ①, but darker brown in older specimens. Margin often has a prominent, yellow growth around the edge ③. Underside has compact, thin gills, which are initially yellow-orange, turning brown with age ②.

TYPICAL FEATURES
The gills are almost always in parallel rows, if the fruits grow out sidewards from the tree.

FLESH: Banded flesh is corky and leathery; rust-brown to dark brown.

DISTRIBUTION: Conifer Mazegill is common and grows in sunny locations on dead conifer trunks and stumps. Also found on fence posts.

EDIBILITY: Not suitable for eating.

SIMILAR SPECIES: Fir Bracket (*Gloeophyllum abietinum*) ④ is rarer and not so brightly coloured. Its gills are less compact.

Root Rot

Heterobasidion annosum

APPEARANCE: Fruiting body 5–20cm wide and a maximum 1.5cm thick; bracket-shaped and very hard and tough. Young mushrooms are light to red-brown ②, with irregular patterning and banding; later covered with a grey-brown or black, bumpy crust ①; margin is always whitish. Fruit layer is decurrent where it joins the tree, and has very narrow, whitish or yellowish pores.

TYPICAL FEATURES
The fruiting body of Root Rot can be easily removed from the tree, as its attaching root is very narrow.

FLESH: Initially white, corky and tough, with a sour smell. Later hard and woody brown.

DISTRIBUTION: This dreaded parasite is found in coniferous woodland. It damages the roots and threatens the life of conifers. The tree roots are turned into a brown, friable mass.

EDIBILITY: Not suitable for consumption.

SIMILAR SPECIES: The Serial Tramete (*Antrodia serialis*) ③ forms flatter fruiting bodies, which can easily be removed from the tree. Whitish with orange-brown edges. It grows in vertical cracks in the bark ④.

Common Bird's Nest
Crucibulum laeve

APPEARANCE: Fruiting body is convex to egg-shaped; young specimens are covered with an orange to ochre-coloured, scaly flocose 'lid', but this falls off with age. The outside of the cup is smooth and felted, initially whitish to ochre-yellow ③, but later brown or black ④. The inside is smooth and creamy or ochre-coloured, with up to 15 white, lentil-shaped sacs of spores ①.

FLESH: Initially soft, but harder with age.

DISTRIBUTION: On rotting plant remains, dead and rotting wood. Rare on soil. Sometimes on cardboard boxes and old sacks.

EDIBILITY: Not edible.

TYPICAL FEATURES
The fungus acquires its characteristic cup or nest shape only after the golden yellow lid has fallen off. The cup reaches a maximum height of 1cm.

SIMILAR SPECIES: Fluted Bird's Nest *(Cyathus striatus)* ② is bigger and always has a furrowed pattern. It only grows on deciduous wood.

Sessile Earthstar
Geastrum fimbriatum

APPEARANCE: The star shape of this fungus develops when the 2–5cm-wide, underground sphere emerges from the ground. This splits the outer layer into as many as 10 whitish-ochre lobes, which unfold into a star pattern. This outer layer then gradually rolls up, releasing the stemless, light brown inner sphere ①. The sphere finally opens up at the top ② and releases the powdery spores.

FLESH: The flesh has almost no smell and is the same colour as the lobes. Initially has a fleshy consistency, but is later dry.

TYPICAL FEATURES
The cream-coloured Sessile Earthstar is easy to identify by the thin, outer layer and the inner sphere, which bursts open at the top.

DISTRIBUTION: Found in clusters, mainly in fallen needles in spruce woodland, on dry, limestone soils.

EDIBILITY: Not poisonous but not suitable for consumption.

SIMILAR SPECIES: Rosy Earthstar *(Geastrum rufescens)* ④ is larger and grows in deciduous and coniferous woods. Flesh turns a reddish colour ③.

Mosaic Puffball

Calvatia utriformis

APPEARANCE: The grey-white to beige fruiting body is 10–15cm wide. It has a low, stemless sac shape ① and is covered in white, floury spikes that disappear with age. The surface of the fruiting body then develops a cracked, mosaic patterning ③. When mature, the cap section of the fungus turns brown and bursts open, releasing the spores ②.

FLESH: When young, the spore mass is separate from the stem flesh, and is white and relatively firm, with a very faint smell ④. It later becomes a yellowish green and eventually turns brown.

TYPICAL FEATURES
The Mosaic Puffball is rare and has an obvious division between the stem and the cap. It grows on fertilised meadows and at woodland edges.

DISTRIBUTION: The fruiting body is mainly found in mountains, mostly on dry meadows, heathland, and woodland edges, less frequently in sparse woodland and never in dense woods.

EDIBILITY: If the inside is firm and white, the fungus is edible.

Stalked Puffball

Calvatia excipuliformis

APPEARANCE: The fruiting body is 8–12cm high and can take a variety of shapes. It normally has a round head on a stalk ①, but is occasionally sac-shaped or pear-shaped. The head can be up to 10cm wide and beneath it, the mushroom is often wrinkly ②. Its white to greyish-white surface becomes brown with age and is covered in 1mm long 'spikes', which can be washed off easily. When ripe, the head disintegrates and releases the brown spores ④.

TYPICAL FEATURES
Stalked Puffball is taller than it is broad, and almost always has a distinct stem.

FLESH: The flesh of young fruiting bodies is white and relatively firm. It has no real taste or smell. The inside of the head is olive-brown and powdery when ripe ③.

DISTRIBUTION: Relatively rare and grows in deciduous and coniferous woodland, only occasionally in meadows. Almost always found in large clusters.

EDIBILITY: When the flesh is white and firm, it is edible. The skin remains tough, even when thoroughly cooked.

159

Brown Puffball
Bovista nigrescens

APPEARANCE: The spherical fruiting body of young specimens closely resembles an egg ①. It is 3–6cm wide. The exterior skin is whitish turning brown when pressed. When ripe, this skin peels off like pieces of eggshell, and the dark brown or black inner layer becomes visible ②. An opening forms at the top ③, through which the spores are released.

TYPICAL FEATURES
The inside of the egg-shaped fruiting body is completely filled with brown spores when mature.

FLESH: The soft flesh is initially grey-white and has little smell or taste. When ripe, the flesh disintegrates to form brown or dark purplish-brown powdery spores ④.

DISTRIBUTION: Mainly found in mountains, rare in lowland areas. Grows in deciduous woodland, on woodland edges and beside footpaths, in meadows, dry grassland and on fallow land.

EDIBILITY: Only young fruiting bodies, which are still pale on the inside, are edible.

Giant Puffball
Langermannia gigantea

APPEARANCE: The irregular, spherical fruiting body is wrinkled at the base ②. These fungi can grow to over 50cm in diameter, and weigh several kilograms. The outer skin is initially leathery, smooth and white ①, but turns brown with age. When ripe, this layer peels off in flakes. The ripe fruiting body is full of powdery spores ④ and suddenly bursts open, detaching from the root-like hyphae.

TYPICAL FEATURES
The enormous size and the smooth, subsequently flaky skin, make the Giant Puffball easy to identify.

FLESH: Flesh of young specimens is firm ③ and has a pleasant mushroom-like smell and taste; later becomes cottony, spongy and dry, with an unpleasant smell and an olive or brown colour.

DISTRIBUTION: Grows in nitrogen-rich meadows, heathland and light woodland edges. Particularly common after wet summers.

EDIBILITY: Young Giant Puffballs with white flesh can be prepared by removing the skin, slicing the flesh, and frying in breadcrumbs.

Spiny Puffball
Lycoperdon echinatum

APPEARANCE: Young fruiting bodies are spherical ②, becoming slightly pear-shaped as they get older ③. The brown surface is covered in tough excrescences up to 5mm long, which are called warts or spines ①. After these spines fall off or are washed off, the inner layer, which is divided into patches, is revealed – this has a regular, reticulated pattern ④. When the 5cm-tall fruiting bodies are ripe, a small opening forms at the top, and spores are released in small clouds whenever the fruit is pressed.

TYPICAL FEATURES
The surface of the Spiny Puffball has dark, prominent patterning once the long spines have fallen off.

FLESH: Initially white, later purple or dark brown, partly divided into cell-like compartments. Finally disintegrates into brown spores.

DISTRIBUTION: Normally on limestone soils under beeches; sometimes under oaks or hornbeams, but extremely rare in coniferous woods.

EDIBILITY: Not recommended for eating.

Common Puffball
Lycoperdon perlatum

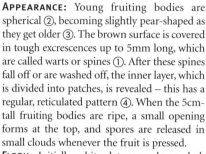

APPEARANCE: Fruiting body is an inverted bottle or club-shape ①; 6–8cm high. Surface is covered with soft, conical points, about 2mm long ②, which fall off easily and leave behind a reticulated patterning. Colour is initially white, but greyish-brown when mature. Eventually bursts open, puffing out clouds of spores when touched.

TYPICAL FEATURES
Common Puffball always has a prominent head and stalk. The flesh has an aroma of radishes.

FLESH: Young specimens are white and fairly firm; as they age, they first become squashy and olive-brown, and end up dry.

DISTRIBUTION: In deciduous and coniferous woodland; very common and always in clusters.

EDIBILITY: Young specimens of Common Puffball are edible when the flesh is still white and can be sliced.

SIMILAR SPECIES: Stump Puffball (*Lycoperdon pyriforme*) ④ is tough and brown from the outset, bald or with fine grains, and has an unpleasant smell of coal gas. White, root-like hyphae hang from the base ③.

162

163

Common Earthball

Scleroderma citrinum ☠

APPEARANCE: Fruiting body is tuberous or club-shaped and up to 10cm wide; reminiscent of a potato. Outer skin is whitish-yellow to ochraceous, with rough, fissured, scaly patterning ①. When ripe it bursts open ④ and releases the spores from a ragged opening.

FLESH: Flesh of young specimens has lilac-grey spotting. When ripe, the spore mass is black with whitish veins. Finally disintegrates into powder with a piercing metallic aroma.

DISTRIBUTION: On sandy or marshy soils in coniferous woodland.

TYPICAL FEATURES
In cross-section ① the 5mm-thick, rough, whitish shell and the ripe, black mass of pores are visible.

EDIBILITY: Common Earthballs used to be used to create fake truffle dishes, though it is now known that the mushroom is mildly poisonous. Rapidly causes nausea, vomiting and loss of consciousness.

SIMILAR SPECIES: Scaly Earthball (*Scleroderma verrucosum*) ② is less common, softer and paler. It has a small stem ③.

Devil's Fingers

Clathrus archeri

APPEARANCE: First a sort of egg-sac develops, a white-grey, spherical shell, surrounded by a membrane and a thin outer skin ②. Inside this egg, the reddish-brown structure of the mushroom is already formed ④. The egg bursts open and the pale red fruiting body emerges ③. When fully open and extended, it has between three and six pointed 'fingers', each 5–12cm long. These have are a bright red colour and look like the tentacles of a squid ①.

FLESH: Flesh is spongy and porous.

TYPICAL FEATURES
The rough surface of the 'fingers' is covered with olive-brown spore sacs, which smell of rotting flesh. The spores are dispersed by flies.

DISTRIBUTION: Native to south-east Asia and Australia and introduced accidentally into Europe 50 years ago, probably in fleeces sent to Europe for turning into wool. It is now spreading rapidly, the spores being dispersed by flies, attracted by the noxious smell. Grows in deciduous woodland and in parks.

EDIBILITY: This curious looking fungus is not edible, because it smells of rotting flesh.

Dog Stinkhorn

Mutinus caninus

APPEARANCE: The rod-shaped fruiting bodies are 8–12cm high and develop in white 'eggs', about 2–3cm large ③. The tips of the fruiting bodies burst through the shell of the eggs ② and the remnants remain, clinging to the base ④. The shafts are hollow, whitish and covered in wide pits. They are tipped with a 1–1.5cm long point, covered in an olive-green, slimy mass of spores ①. The fruiting body is initially vertical, but slowly bends over and eventually lies on the ground. The spores are dispersed by insects.

TYPICAL FEATURES
The dark, olive-green spike of the thick fruiting body is pierced at the tip.

FLESH: The whitish to ochre-yellow interior of the fruiting body is hollow or very porous. Smells unpleasantly of cat excrement.

DISTRIBUTION: Dog Stinkhorn is not very common, but can be found in large clusters in deciduous and coniferous woodland with a lot of rich top-soil or humus.

EDIBILITY: Inedible.

Stinkhorn

Phallus impudicus

APPEARANCE: The fungus begins as an egg-shaped mass in the ground ②. It then develops fruiting bodies up to 20cm high. The dark, olive-green tips ④ are covered in a mucous and are clearly separate from the 3cm-thick hollow stem with its honeycombed pits ③. The spores are located on the sticky tip and are dispersed by flies, which are attracted by the smell of rotting flesh ①. The membranous remnants of the egg-like universal veil remain around the stem base, as in the Death Cap (p62).

TYPICAL FEATURES
Stinkhorn has a very unpleasant smell of rotting flesh that can be detected from a distance of 20m.

FLESH: Smells of radish at first. Ripe fruiting bodies smell of rotting flesh. The stem is very porous.

DISTRIBUTION: Found throughout Europe, Asia and North America, mainly in deciduous and coniferous woodland with a lot of topsoil.

EDIBILITY: Inedible when ripe. The 'eggs' are edible.

Cauliflower Fungus

Sparassis crispa

APPEARANCE: The yellow or pale cream-coloured fruiting body ① can be over 30cm in diameter and consists of sinuous, foliate, wrinkled layers ②.

FLESH: The flesh has a mild flavour and pleasant smell. Older specimens have a bitter taste.

DISTRIBUTION: A root parasite in coniferous woodland, mainly under pines.

EDIBILITY: Young specimens are valued by mushroom pickers as a tasty edible mushroom. Small creatures, sand and needles become

TYPICAL FEATURES
Cauliflower Fungus is unmistakeable due to its spongy, or lettuce-like appearance, and its diameter of up to 30cm.

trapped in between the leaves and can be hard to remove.

SIMILAR SPECIES: The inedible Umbrella Polypore *(Polyporus umbellatus)* ④ grows on the roots of old deciduous trees, such as oaks and beeches. It consists of numerous small, stemmed caps ③.

Ice Coral

Hericium flagellum

APPEARANCE: The fruiting body is 30–40cm tall and looks like coral ①. The base of the fungus is attached to the trunk of a tree, and it develops several forking branches, each about 1cm thick, and bearing 2cm-long 'spines' ②. These spines are always white, but the branches of older specimens can turn flesh-pink or ochre-coloured.

TYPICAL FEATURES
Ice Coral is very distinctive and grows on fir trees that are still standing, but which have lost their bark.

FLESH: White and soft, but often tough. Taste is mild, but smell is unpleasant.

DISTRIBUTION: Found in woodland on old firs and spruces. Although it is widespread, it is threatened because rotting wood is now quickly removed. Has been seen in the New Forest.

EDIBILITY: The fungus is not poisonous, but should not be collected because it is endangered.

SIMILAR SPECIES: The inedible Coral Tooth *(Hericium coralloides)* ④ has regular spines, arranged like a comb ③.

Golden Coral Fungus
Ramaria largentii

APPEARANCE: Fruiting bodies are up to 12cm tall and 6–15cm wide; yellow to orange-yellow in colour, and densely branched like coral ①. Short trunk is a slightly paler yellow than the forked, pointed branches ④. Fruiting bodies become paler with age.

FLESH: Flesh is mild and brittle. In the trunk it is a dirty white, but elsewhere it is yellowish. Flesh of older specimens becomes tough.

DISTRIBUTION: On patches of moss in coniferous forests in mountain regions.

TYPICAL FEATURES
Golden Coral Fungus is easily identified by its bright yellow colour and its distribution in coniferous woodland.

EDIBILITY: Not poisonous, but a very poor quality edibility. Some poisonous species are very similar in appearance so this mushroom should not be collected except by experts.

SIMILAR SPECIES: The bitter-tasting Pale Yellow Coral (*Ramaria flavescens*) ② belongs to a group of yellow corals that mainly grow in deciduous woodland. The branch tips are slightly serrated ③.

Yellow-tipped Coral
Ramaria formosa

APPEARANCE: Fruiting body is 8–12cm high and similarly wide. It is very branched, like coral ①. The salmon-coloured branches end in two or three lemon-yellow spines ②, which become salmony-orange with age. The fungus forms a short, white trunk where it joins the tree.

FLESH: White to pale pink. The points have an acrid taste. Otherwise mild or slightly sour. Almost without smell.

TYPICAL FEATURES
Young specimens can be identified by the salmon-coloured branches, with their yellow tips ②.

DISTRIBUTION: This eye-catching coral can be found in deciduous woodland, mainly on limestone soils, under beeches.

EDIBILITY: Mildly poisonous and can cause nausea and serious diarrhoea. Even edible species of coral fungus can cause stomach upsets, so their consumption is not recommended.

SIMILAR SPECIES: The inedible Flaccid Coral (*Ramaria flaccida*) ④ has thin, vertical, ochre-yellow branches. The Rosso Coral (*Ramaria botrytis*) has branch tips which are pinkish-violet in colour ③.

Yellow Stagshorn
Calocera viscosa

APPEARANCE: The fruiting bodies are 4–8cm high and have a white, felt-like base. This base forms root-like extensions ②, which can extend into the wood to a depth of 25cm. The shape is initially clubbed, but quickly becomes bushy and coral-like ①. It is golden-yellow to orange-red in colour ③.

TYPICAL FEATURES
Yellow Stagshorn has elastic, tough and flexible branches, with gelatinous, sticky tips.

FLESH: The flexible, elastic, tough flesh becomes as hard as bone when dry. It has no smell or taste.

DISTRIBUTION: Grows on dead coniferous wood. Often grows on spruce stumps that are covered in moss.

EDIBILITY: Not poisonous, but cannot be eaten, because of its tough texture.

SIMILAR SPECIES: Small Stagshorn (*Calocera cornea*) ④ is closely related, but has few, if any, branches. It only grows on dead deciduous wood.

Crested Coral
Clavulina coralloides

APPEARANCE: This coral fungus is 2–10cm large and initially snow-white, with very dense branches ①; turning grey when mature, when only the tooth-like points on the comb-like branches remain white ②.

FLESH: The whitish flesh has a soft and slightly brittle texture. It has a mild, occasionally slightly bitter, taste and almost no aroma.

TYPICAL FEATURES
This coral fungi is pure white or slightly grey. It is characterised by the comb-like branch tips, which are often very flat, but which always remain white.

DISTRIBUTION: Common and grows individually or in groups on needles beneath spruces. Also found in deciduous woodland.

EDIBILITY: Not edible.

SIMILAR SPECIES: The Grey Coral (*Clavulina cinerea*) ④, is grey from the outset and is often wrinkled. Its branches are very densely forked, but blunt. Purple Coral (*Clavulina amethystina*) ③ has very brittle branches and becomes brown with age; it has a greyish-yellow colour when dry.

173

Jelly Ear
Auricularia auricula-judae

APPEARANCE: Fruiting body is initially an irregular cup-shape ③, but later becomes the shape of an ear ①, with a small, stem-like protuberance which attaches to the wood ④. Surface is brown or black, with a silky and slightly grainy texture. The underside is purplish-brown, with ribbed, veined wrinkles.
FLESH: Has no smell or taste. Thin and translucent, and brown and gelatinous. When dry it is as hard as bone and brown or black.

TYPICAL FEATURES
The purplish-brown, wrinkled underside may have a frosted appearance when the spores are released.

DISTRIBUTION: Found year round, but mainly in winter. In dense patches on old or dead elder wood or other deciduous wood.
EDIBILITY: A related species is used in Chinese cooking. Used as a salad mushroom or soup flavouring in Europe. Dried mushrooms swell when placed in water.
SIMILAR SPECIES: Tripe Fungus (*Auricularia mesenterica*) ② is inedible and has shaggy hairs on the surface.

Salmon Salad
Tremiscus helvelloides

APPEARANCE: Fruiting body is initially spatulate ②, but becomes tubular or funnel-shaped when ripe; 6–8cm high; one side has a crevice running its entire length ③; colour is a variety of red shades, from orange-red to brownish-red or salmon-red ①; the base of the stem is always paler and almost whitish. The outside is hairless and often has a slightly purple frosting; initially smooth ④, but later slightly wrinkled.
FLESH: The translucent, gelatinous, elastic flesh becomes as hard as bone when dried. It is

TYPICAL FEATURES
Salmon Salad is easy to recognise due to the reddish fruiting bodies, which are funnel-shaped with a split down one side.

paler than the outside of the mushroom and has a watery taste.
DISTRIBUTION: Shady, damp locations in coniferous woodland, beside forest footpaths, in wood stores and under bushes. Found both on the ground and on dead wood, mainly in clusters; most common in mountain regions.
EDIBILITY: Edible, but not very tasty. Young fruiting bodies can be eaten raw, something that is only true for very few species of fungus.

False Morel
Turban Fungus | *Gyromitra esculenta* ☠

APPEARANCE: The fruiting body has distinct stem and cap sections. The cap is dark brown or red-brown and has a lobed, sinuous surface and a brain-like structure ②. It can be 8–10cm wide and is not hollow. Below the cap, the whitish stem is very bent ① and has vertical folds; it is partly hollow ③.

FLESH: A pleasant aromatic smell; firm, especially in the stem. Flesh of cap is more brittle.

DISTRIBUTION: Mainly in spruce woodland, on sandy soils; normally in droves.

TYPICAL FEATURES

False Morel has a characteristic brain-like, sinuous structure. It is not honeycomb-like. Unlike the edible Morel (p178), the cap is not hollow.

EDIBILITY: Contains the poison gyromitrin, which is not rendered harmless by cooking, as is often reported. Frequent consumption has caused many deaths.

SIMILAR SPECIES: Pouched False Morel (*Gyromitra infula*) ④ is also thought to be poisonous and also grows in coniferous woodland, but typically grows in autumn.

White Saddle ⚑
Helvella crispa

APPEARANCE: The fruiting body is whitish to light ochre-brown and is 5–15cm high ①. The cap has very folded lobes ②, and is often saddle-shaped on younger specimens ③. The whitish stem also has an irregular shape and deep vertical grooves ④.

FLESH: The whitish, elastic, tough flesh of young fruiting bodies has no taste and almost no smell. Older specimens smell slightly sweet. The texture of the cap is slightly brittle, and the stem flesh is slightly tough.

TYPICAL FEATURES

White Saddle is relatively large and is characterised by the whitish-grey or grey-brown colour of the cap, and the lobed, grooved shape of the fruiting body.

DISTRIBUTION: Common saprophytic fungus in some areas, mainly in deciduous and mixed woodland, in parks, beside footpaths and on areas of ground covered in grass and moss.

EDIBILITY: Edible when thoroughly cooked, but consumption is not recommended. Older specimens can be very poisonous.

Conical Morel
Morchella conica

APPEARANCE: The cap is conical and distinct from the stem. It has many hollow indentations, demarcated by horizontal and vertical ridges ①. The fruiting body can be 15cm high and the cap is normally longer than the stem. The stem has a hollow centre ④ and has a whitish, cream or grey colour.

FLESH: The whitish flesh has no smell, but a mild and pleasant taste.

DISTRIBUTION: Particularly in alluvial forests, coniferous woodland, on grassy woodland edges and beside footpaths. Found in spring.

EDIBILITY: Conical Morel is a deliciously edible mushroom.

SIMILAR SPECIES: Semifree Morel (*Mitrophora semilibera*) ② is also edible but smaller, and often has a bell-shaped, overhanging cap ③.

TYPICAL FEATURES
When identifying the Conical Morel, it is important to look for the vertical ridges in the dark, almost black cap. It is only found in spring to early summer.

Common Morel
Morchella esculenta

APPEARANCE: Up to 6cm high, with an elongated or egg-shaped cap ①, which can sometimes also be rounded. Surface can be honey-coloured, ochre-yellow or light brown, and sometimes even grey-brown; it has ridges which run in an irregular pattern, forming a honeycomb effect of chambers ④. The smooth fruit layer is located on the inside ③. The stem is whitish or grey, up to 9cm long, and is often bent, with an irregular shape ①. Cap and stem are hollow ②.

FLESH: The brittle flesh is white and has no smell. It tastes mild.

TYPICAL FEATURES
The irregular pattern of the ridges on the elongated cap, which forms a honeycomb-like surface structure, is very characteristic of the Morel.

DISTRIBUTION: On damp soils in mixed woodland, alluvial forests and parks. Sometimes also in gardens, hidden in the grass.

EDIBILITY: Morels are considered to be among the finest edible fungi. Older specimens can cause stomach upsets and should be avoided.

179

Scarlet Elf Cup

Sarcoscypha austriaca

APPEARANCE: The fruiting bodies of young specimens are cup-shaped and have a short stem ②. Later broadens to form a 2–5cm wide dish, with a very inrolled margin. The inside of the fruiting body is a bright vermillion red ①; the skin on the outside is paler.

FLESH: The tough flesh has no distinct smell or taste.

DISTRIBUTION: Eye-catching but relatively rare. Appears during the spring thaw. Saprophytic and grows on dead deciduous wood.

EDIBILITY: Scarlet Elf Cup is edible, but does not have a high yield and should be protected.

SIMILAR SPECIES: Orange Peel Fungus *(Aleuria aurantia)* ④ is bright orange in colour. Fully grown specimens have a lobed and very broad shape ③. It is not edible.

> **TYPICAL FEATURES**
> The bright red colour of the cup-shaped fruiting body makes the Scarlet Elf Cup easy to identify.

Violet Crowncup

Sarcosphaera crassa

APPEARANCE: This striking mushroom forms a fruiting body which is 5–10cm wide, and very rarely up to 15cm in diameter. It develops underground as a hollow ball ②. When it emerges from the soil, the shell breaks open ③, forming its typical crown shape ①. The smooth inside of the open fruiting body is brownish or white and often takes on a purple shade. As it gets older, it often turns dark brown ④.

FLESH: The white, very brittle flesh only has a very faint smell.

> **TYPICAL FEATURES**
> Violet Crowncup is hard to confuse with any other species, because of its fruiting body, which bursts open to form a star or crown shape.

DISTRIBUTION: Fruiting bodies are rarely seen. It is a saprophytic fungus, growing on limestone soils in coniferous and deciduous woodland.

EDIBILITY: Poisonous and causes serious stomach upsets. In some countries it is collected in large quantities and eaten, but consumption is definitely not recommended.

Candlesnuff Fungus

Xylaria hypoxylon

APPEARANCE: The fruiting body is 2–4cm high and greyish-black in colour. It is elongated or club-shaped, and often branches at the top to form antlers ②. Often appears very flat, particularly around the antler-like branches ③. The surface is warty and normally covered in white spore dust ①.

FLESH: The flesh is soft and elastic to woody. It does not really have any smell or flavour.

DISTRIBUTION: Often found on dead deciduous wood, mainly on the stumps of beeches; less common on coniferous wood.

EDIBILITY: Not edible.

SIMILAR SPECIES: Dead Man's Fingers *(Xylaria polymorpha)* ④ is also inedible. It grows 4–8cm high, and can have a variety of shapes, but the fruiting body is always black.

TYPICAL FEATURES

The fruiting body is tough and club-shaped, with antler-like branches. The surface is rough and wart-like, and the branches do not break, even when bent double.

Bay Cup

Peziza badia

APPEARANCE: The fruiting body is a lush chestnut brown, and is the shape of a bowl or shell ①. It is stemless and can reach a diameter of 2–5cm. The inside and outside of the cup have the same colour.

FLESH: The brittle flesh has almost no smell, and very little flavour.

DISTRIBUTION: A relatively common saprophytic fungus in coniferous and deciduous woodland. Grows directly on the woodland floor, where it lives on dead plant remains.

EDIBILITY: The Bay Cup is not edible.

TYPICAL FEATURES

Damaged parts of the fruiting body release a colourless, watery juice, which quickly turns brown when exposed to air.

SIMILAR SPECIES: The fruiting bodies of Blistered Cup *(Peziza vesiculosa)* ④ are up to 10cm wide and have a bulbous outer surface with large grains. The Yellowing Cup *(Peziza succosa)* ② grows on damp moorland soil and releases a milky juice, which quickly turns yellow. Purple Cup *(Peziza moseri)* ③ grows on burnt ground in coniferous woodland and has a purplish colour.

183

Lemon Disco

Bisporella citrina

APPEARANCE: Often forms dense coverings on dead wood ①. The bright lemon-yellow fruiting bodies are rounded or plate-like ②. They form on a short stem and are a maximum of 3mm across.

FLESH: Lemon yellow throughout. Has no smell or taste.

DISTRIBUTION: In dense coverings on dead deciduous wood, mainly on beech wood.

EDIBILITY: This tiny fungus is not suitable for eating.

TYPICAL FEATURES
This bright yellow fungus is normally found on beech wood from which the bark has been removed and on fallen branches.

SIMILAR SPECIES: Beech Jellydisc *(Neobulgaria pura)* ③ also grows on lying beech trunks, but is much larger, with a diameter of 2–4cm. Witches' Butter *(Fuligo septica)* ④ is bright yellow, and does not form individual pustules; the fruit body is a slimy, spongy mass, which can be over 20cm across.

Common Jelly Spot

Dacryomyces stillatus

APPEARANCE: This fungus has two fruiting forms. One is a fruit body which is only 2–3mm large and golden yellow or orange-yellow in colour ①. It has the form of a small, irregular pustule. Surface is slightly folded and sinuous. The second form is more common ④ and is even smaller. Red-orange and convex, melting away when mature.

FLESH: The gelatinous, soft and watery flesh has an insipid smell and flavour.

DISTRIBUTION: Year round on cut coniferous logs; less common on deciduous woods.

EDIBILITY: The Common Jelly Spot is not edible.

TYPICAL FEATURES
The spot-like fruiting bodies mainly form on the cut surfaces of deciduous and coniferous stumps, or on trunks with the bark removed.

SIMILAR SPECIES: Yellow Brain *(Tremella mesenterica)* ② is also bright yellow and grows on the branches of dead wood. The gelatinous fruiting body is sinuous and brain-like, but can be over 10cm wide. The pustules of the very common Coral Spot *(Nectria cinnabarina)* are 2–3mm wide ③ and are found on both dead and living deciduous wood.

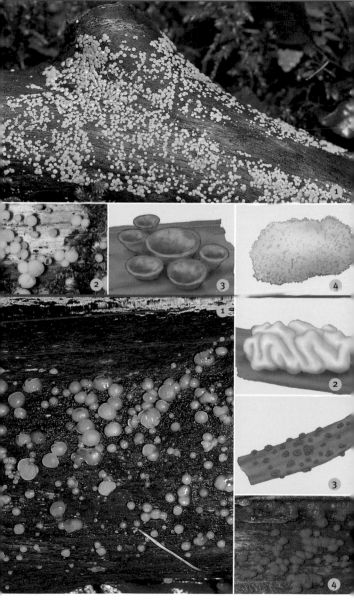

Index

186

189

Acknowledgements

Cover picture: Bay Bolete; small picture on the left: Yellow Stagshorn, Milking Bonnet, Hollow Bolete
Pages 6/7: Sheathed Woodtuft
Pages 24/25: Fly Agaric

Bahr: Pages 9 top; blickwinkel: (Aßmann) Pages 177 bot. r., (Behlert) Pages 17 bot., 147 top, 181 top r., (Mosebach) Pages 27 top, 93 b.r., (Perseke) Pages U 1 top l., 37 bot. l., 63 top l., 101 top l., 121 b.r., 149 bot., b.r., 151 bot., 169 top l., 173 top l., (Schaber) Pages 3 bot., (Schaub) Pages 24/25, (Schröer) Pages 53 top l., (Schütz) Pages U 2 l. mi., 147 top l., (Tomm) Pages 153 top, 159 b.r.;
Garnweidner: Pages U 1 top mi., U 2 l. top, r. top, r, mi. l. bot , 2, 3 top, 8 top, bot., 9 mi., bot., 11 r. top, r. bot., 12, 13, 14 bot., 16, 17 mi., 18, 19, 21 bot., 22, 23, 27 top l., bot., b.r., 29, 31, 33 top l., b.r., 35, 37 top l., 39 top, top l., bot., 41 top, bot , b r , 43 top, bot , 45 top, top l., b.r., 47 top, top l., 49 top bot., 51 bot., 53 b.r., 55, 57, 59 top, bot., b.r., 61 bot., 63 top, bot. l., 65 b.r., 67 top, top l., 69 top r., bot. l., b.r., 71 top, bot., b.r., 73 top, bot. l., b.r., 75 top, top l., bot., 77, 79 top, top l., b.r., 81 top, bot., b.r., 83, 85, 87 bot. l., top r., b.r., 89 top l., top r., b.r., 91 top, bot., b.r., 93 top l., bot., 95, 97, 99 top l., top r., bot. l., 101 top l., top r., b.r., 103 top r., bot. li, 107 top, top l., bot., 109, 111 top l., b.r., 113, 115, 117, 119, 121 top, top l., bot., 123 top l., top r., bot., 125 top, bot., b.r., 127 top, bot., b.r., 129, 131 top, top l., bot. l., 133 top, top l., bot., 135, 137, 139 top l., top r., bot., 141, 143 top, bot., 145 top, bot., 147 b.r., 149 top r., 151 top r., b.r., 153 top l., 155 top r., bot., 157 top r. bot , b m, 159 top, top l., bot. l., 161 top l., b.r., 163 top, bot., b.r., 165 top l , 169 top r., bot. l., b.r., 171, 173 b.r., 175 top l., 177 top l., bot. l., 179 top, bot. l., 181 b.r., 183 top, b.r., 185 top l., bot. l., b.r., U 4 r.;
Grünert: Page 111;
Hofrichter: Pages U 1 (large picture), 11 l. top, 33 bot., 49 b.r., 53 top r., 99 b.r., 143 top l., 147 bot., 151 top l.;
König: Page 8 mi., 21 top, 61 top l., 67 bot. l., 69 top l., 73 top l., 103 b.r., 163 top l., 167 top l., 173 bot. l., 179 b.r.;
Laux: Pages U 1 top r., U 2 r. bot., 33 top r., 37 b.r., 39 b.r., 41 top l., 43 top l., b.r., 45 bot. l., 47 bot. l., b.r., 51 top l., top r., b.r., 53 top r., 59 top l., 61 b.r., 63 b.r., 65 top l., top r., 67 b.r., 71 b.r., 75 b.r., 79 b.r., 81 top l., 87 top l., top r., 89 bot. l., 91 top l., 103 top l., top r., 105 b.r., 107 b.r., 111 top r., 123 b.r., 125 top l., 127 top l., 131 b.r., 133 b.r., 139 b.r., 143 b.r., 145 topli, b.r., 149 top l., 153 bot. l., 155 top l., b.r., 157 top l., 161 top, bot., b.r., 173 top r., 175 bot. l., b.r., 177 top r., 179 top l., 181 top l., bot. l., 183 bot. l., 185 top;
Reinhard: Pages 11 l. bot., 15, 17 top, 20, 37 top, 49 top l., 61 top, 65 bot. l., 93 top, 103 bot., 161 bot., 165 bot. l., 167 t., bot. l., b.r., 183 top l., U 4 mi.;
Walz: Pages 6/7, 14 t.

Acknowledgements

The dates and facts in this guide have been researched and checked with great care. No guarantee can, however, be given, and the publisher accepts no liability for damage to people, property or assets.

The translators would like to acknowledge the assistance of Elizabeth Holden of the British Mycological society in providing the latest English nomenclature.

This edition first published in 2006 by New Holland Publishers (UK) Ltd
London • Cape Town • Sydney • Auckland
10 9 8 7 6 5 4 3 2
www.newhollandpublishers.com
Garfield House, 86–88 Edgware Road, London, W2 2EA, UK

Copyright © 2006 in translation: New Holland Publishers (UK) Ltd

ISBN 978 1 84537 474 7

Publishing Manager: Jo Hemmings
Senior Editor: Kate Michell
Editor: Kate Parker
Translator: American Pie, London and California

© 2005 GRÄFE UND UNZER VERLAG GmbH, Munich

Series editor: Steffen Haselbach
Desk editor: Anita Zellner
Copy editing: Dr. Michael Eppinger, Dr. Helga Hofmann
Design: independent Medien-Design
Layout: H. Bornemann Design
Illustrations: Peter Braun, atelier amAldi
Typesetters: Filmsatz Schröter, Munich
Production: Petra Roth
Repro: Penta, Munich
Printing: Appl, Wemding
Binding: Auer, Donauwörth
Printed in Germany